Citrus

A GUIDE TO ORGANIC MANAGEMENT, PROPAGATION, PRUNING, PEST CONTROL AND HARVESTING

ALLEN GILBERT

HYLAND HOUSE

First published in Australia in 2007 by
Hyland House Publishing Pty Ltd
Melbourne, Australia
www.hylandhouse.com.au

Reprinted 2010

Copyright © Allen Gilbert 2007

This book is copyright. Apart from any fair dealing
for the purposes of private study, research, criticism
or review, as permitted under the Copyright Act, no
part may be reproduced bay any process without
written permission. Enquiries should be addressed
to the publisher.

National Library of Australia
Cataloguing-in-publication data:

> Gilbert, Allen.
> Citrus : a guide to organic management, propagation,
> pruning, pest control and harvesting.
>
> Bibliography.
> Includes index.
> ISBN 9781864471038 (pbk.).
>
> 1. Organic gardening - Australia - Handbooks, manuals, etc.
> 2. Citrus - Australia. I. Title.
>
> 635.987

Edited by Nerissa Greenfield
Design by Rob Cowpe Design
Printed by Everbest, China

foreword

Let's be brutally honest and openly celebrate the publication of an outstanding, much-needed book. After all, gardeners and horticulturists have been impatiently waiting decades for it. This is about Australia's national obsession – citrus trees, and how to grow them successfully.

Listen to any radio gardening talk-back programme in any part of Australia at any time of the year, and before long up comes the inevitable query about a lemon, orange, mandarin, grapefruit, lime or any other kind of citrus tree. Perhaps this is because almost every backyard features one or more of them, particularly lemon trees. And here at last is a book with all the answers.

Allen Gilbert is one of those rare people who is not only a leading horticulturist but also one of Australia's most experienced, practical, down-to-earth gardeners. To me his words of wisdom are pure poetry. Once again he's excelled himself with this brilliant, beautifully written and illustrated book.

Allen covers everything we all want to know about every kind of citrus. He goes deeply into the history of these valuable, highly productive plants. We discover the origins of different species of citrus and how they were distributed around the world and especially Australia. This alone makes fascinating reading but is only a tiny part of this wonderful book.

The amount of research and knowledge contained within these pages is truly mind-boggling. There is detailed information about every kind of citrus including some unusual species.

Allen writes with the authority of a practical bloke who's been there, done that. He tells us where to plant and how to manage citrus trees for maximum production and good health. We learn about pruning techniques – with his unique and successful methods of pruning lemon trees, how to grow citrus as hedges or fans that take up little space. We can discover tricks for growing citrus plants in tubs, how to propagate from cuttings or seed, the technique of transplanting established trees and the best times to do the job.

Here is a never-ending book packed with sheer knowledge. There are full, highly informative sections on pests and how to control them safely; on diseases and how they are best combated or avoided, and there are plenty of superb photographs for easy, rapid identification.

And of course Allen interprets the messages that citrus leaves send us about a host of nutritional and other disorders with even more illustrations, so we know exactly what can go wrong with our own backyard trees.

This is, without question, the best and most informative book on citrus growing and cultivation I have ever read. It's a bloomin' inspiration!

Peter Cundall

contents

Foreword *by Peter Cundall*	iii
Acknowledgments	vii
1 Citrus: Origins and Spread	1
2 Citrus Species and Cultivars	12
Sweet oranges and other sweet citrus	16
Lemons and other sour citrus fruit	25
Grapefruit and other large citrus fruit	36
Citrus grown for their leaves	39
Australian citrus species	40
3 Propagation of Citrus	42
Propagating citrus by seed	42
Aerial layering	44
Propagation from cuttings	45
Budding and grafting	47
Equipment	48
Rootstocks	49
Scions	52
Budding	53

4	Managing Citrus Trees	64
	Open-grown citrus trees	64
	Citrus in pots	74
	An organic citrus grove	78
5	Pruning and Training Citrus	82
	Pruning and training equipment	83
	Skeletonising citrus trees	84
	Pruning to a stump	85
	Pruning established trees	85
	A new method of pruning lemons	86
	Pruning young trees	86
	Topiary shapes	87
	Training citrus trees as espalier	87
6	Pest and Disease Control of Citrus	91
	Quick guide to citrus problems	91
	Insect and animal pests	98
	Diseases of citrus trees	116
	Non-pathological disorders of citrus	126
	Nematodes, citrus nematode	130
	Nutritional deficiencies of citrus	130
	A–Z of organic pest and disease control	139
7	Harvesting and Storing Citrus Fruit	142
	Harvesting citrus	143
	Storage of citrus	143
Resources		151
References		153
Bibliography		155
Glossary		159
Index		165

acknowledgments

This book would never have been finished in time for the publisher were it not for my wife Laurie Cosgrove. She not only corrects my bad English but assists with research, does all the editing and provides many of the recipes.

Two important people to thank are the first draft readers, both citrus experts and long-term colleagues who I met through the International Plant Propagators Society (IPPS).Neither wished to be named but I thank them wholeheartedly.

The late H.F. Janson, gave permission to use material from his book *Pomona's Harvest* (Timber Press, 1996). Geoff Godden, ex-Victorian Department of Agriculture Citrus Advisor, needs special mention for all his professional advice about citrus during our time as horticultural advisors with the Victorian Department of Agriculture. Denis Crawford checked the scientific names of all the insects mentioned and Ian Pascoe did likewise for the scientific names of all diseases. I am grateful to them both.

I thank the 3CR Collingwood community radio gardening show gang (including many ex-members of the Victorian Department of Agriculture's Garden Advisory Service). Of that group, I must first thank the late Ian Nichols, who was with the show for over 15 years, but also Stephen Ryan, George Edson, Gwen and Rodger Elliot, Pam Vardy, Bruce Hedge, Rosemary Davies, Phillip Hicks, Diana Sargent, Merryl Johnson, Geoff Stockton, John Patrick, Jane Edmanson, Penny Woodward, Brigitte Lyons, Mark Tracey, Greg Moore, the late Don Huxley, Brian Clarke, Dawn Fleming, Bill Molyneux and all the guests on the program.

Dr Steve Sykes has given permission to use material from a lecture entitled 'Breeding with Australian Native Citrus' presented at the International Plant Propagators' Conference, Mildura, 2005. The

International Plant Propagators' Society (IPPS) has been an invaluable resource.

I have been blessed with the friendship of Peter Cundall and cannot thank Peter enough for his support over the years and for offering to write the foreword to this book.

Over the years, many, many people have allowed us to take photographs, and some gardeners have allowed me to carry out budding and grafting, pruning and tree shifting experiments with their precious citrus trees and I am grateful to them. Other people have helped with information or supplied material for the manuscript. Thank you especially to:

    Alexis Goldman for information about preparing limes and lemons
    Anne and Peter Latreille for allowing work on espalier training
    Charlie Turnbull, Bruny Island, for information on citrus pest and disease control methods in China
    Gary and Elaine Dettman for the opportunity to carry out experiments on their property
    Greg Daley, Daley's Nursery for their citrus catalogue list
    Gwen McCrorey, ex-colleague, for recipes from the Victorian Department of Agriculture Journal
    Helen Smyth, Bruny Island, book finder extraordinaire, who found many books and references, some of which were new to me
    Jan Denham, Robert and Damien Ridgewell of Karra Organic Farm
    John Gorr (Elephant) for assisting with Jewish names and words
    Norma Campbell for supplying many delicious citrus recipes
    Paul Croxton and Alan Saunders, Boulevarde Nursery, Mildura
    Tim Henman for Auscitrus information presented at the International Plant Propagators Conference, Mildura, 2005
    Theo and Angela Pitsounis for allowing grafting experiments on their citrus trees
    Nerissa Greenfield for editing
    Peter and Maggi Corrigan
    Eva and John Seles
    Julie and Brian Hurse for photographs of their lemon tree
    Cynthia Carson for sourcing photographs and supplying valuable information

CHAPTER 1

citrus: origins and spread

All species of citrus with which we are familiar today fit into one large genus of plants. The citrus genus belongs to the plant family Rutaceae, containing about 1000 species in about 100 genera (see Glossary for an explanation of plant naming conventions). The family Rutaceae contains many common garden plants including the herb Rue (*Ruta graveolens*), Mexican orange (*Choisya ternata*), Orange jessamine (*Murraya paniculata*) and Australian native plants such as Boronia (*Boronia* spp.), Correa (*Correa* spp.) and Crowea (*Crowea* spp.).

Origins

The citrus genus probably started to evolve millions of years ago before the tectonic plates of the earth shifted to create separate continents, resulting in separate evolutionary patterns and distinctive plant species. Citrus species occur naturally in the tropics or semi-tropics on several continents including Asia, India and Australasia. The Australian species of interest include *C. australasica* (Finger lime), *C. australis* (Wild lime), *C. garrawayae* (Mt White Lime), *C. inodora* (Russell River lime) and *C. glauca* (Desert lime). In some areas of the world introduced citrus trees have naturalised so that they seem to be native.

It seems almost indisputable that most of the modern edible citrus cultivars have their origin in southern China, and from there they have spread to the rest of the world via historic trade routes. The story of cultivated citrus species is fascinating, with the first recorded mention of citrus made in about 500BC by writers in China. The first manuscript on citrus entitled *Chu lu* was written by Han-Yen-Chi in China in 1178AD.[1] Twenty-seven cultivars of sweet, sour mandarin and other citrus types are meticulously described and their cultural requirements given.

As with other species of garden plants originating from China, such as the Peony rose, it is thought that many citrus cultivars were chosen in a process of selection and breeding over hundreds or thousands of years to give the sweet fruits known today. In my travels in China, I have seen many different cultivars of citrus in gardens and market places, with tremendous variation, especially amongst mandarins, cumquats and pummelos. There are, though, very few of the original citrus species growing in the wild in China today.

The ancestors of the sweet fruited citrus were probably plants producing small, very sour, dry or bitter fruit quite possibly almost inedible by today's taste standards. A modern example of this is the Australian citrus species (*Citrus australasica* syn *Microcitrus australasica*), the Finger lime. Aboriginal people harvested fruit but the trees were not cultivated to any great extent. Many of the wild plants produce inedible fruit with little flesh or juice but some produce sweeter and juicier fruit with a large flesh to skin ratio. Others produce sweet, red-fleshed fruit. These and other variations occur within a plant population that has not been selectively bred by humans until very recently.

Although many citrus species can produce parthenocarpic fruit (without seed; see Glossary), some (such as lemons) have the ability to grow a variety of seedlings from just one seed (see Chapter 3). This

Australian Finger lime selections

means that new cultivars can result from propagation by seed, especially where cross-pollination occurs with other citrus species. This ability, together with the ease with which citrus fruit and seed can be transported over long distances and the relative ease with which citrus can be propagated from seed, facilitated the movement of citrus species within China and further afield, resulting in the early evolution of many different citrus cultivars throughout the world. Two examples of this are the common lemon, and the citron. Both spread in very early times to several parts of the world, and then, as time went by, different seedlings developed separately to produce different popular fruiting cultivars of the same species in different countries so that there are many cultivars of the most common citrus species worldwide.

In his book, *1421, the Year China Discovered the World*, Gavin Menzies theorises that during the early 15th century and possibly earlier as well, Chinese ships sailed the seas of the whole world exploring many places including Australia and New Zealand. Chinese explorers had known of the Australian land mass and visited its shores hundreds of years beforehand. During the 15th century voyages (and probably during earlier exploration voyages as well), the Chinese sailors carried citrus fruit with them to prevent scurvy (see Glossary). This was well before European understanding of the role of citrus in preventing scurvy. Chinese ships carried potted citrus plants and citrus seed, and these as well as other plant species were introduced to places they visited or wherever they established small colonies.

Later, European sailors, including Spanish and Portuguese, also contributed to the spread of citrus species. Citrus seed was carried by Columbus on his second voyage during 1493 and were introduced to parts of the New World. English sailors were nicknamed 'Limies' because of their habit of carrying limes or lime juice on board, a name still used occasionally today.

Groves of various citrus species can be found along the historic trade routes of the world and in some areas citrus trees have become naturalised. There are references to oranges in Indian literature going back more than two thousand years and, while some botanists believe that these are separate species, the actual place of origin is still unclear.

It was possibly the Persians who spread citrus, especially the citron, to many parts of the world and they may have been indirectly responsible for introducing the orange to Europe. The earliest orange to be grown in Europe, specifically in Italy and Spain, was the 'Seville' or 'Sour orange',

In many countries, citrus species have developed religious significance. The Esrog or Etrog is a selected citron that plays an important part in the Jewish Feast of the Tabernacles. In Italy there are orange festivals and in Vietnam during New Year festivities citrus is used as a ceremonial tree.

CITRONS

Frederic, Lord Leighton or Baron Leighton of Stretton, the English classicist and first painter to be given the rank of Baron, painted a famous painting 'Old Damascus: Jews' Quarter, or Gathering Citrons, 1873–74' depicting women gathering citrons in a walled garden in Damascus.

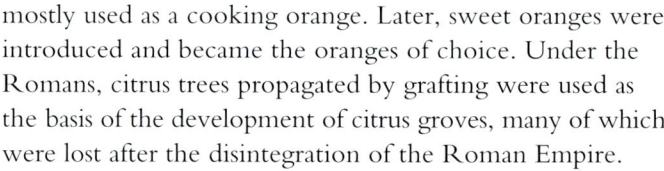

mostly used as a cooking orange. Later, sweet oranges were introduced and became the oranges of choice. Under the Romans, citrus trees propagated by grafting were used as the basis of the development of citrus groves, many of which were lost after the disintegration of the Roman Empire.

Introduced citrus species came to be particularly valued in colder European countries, particularly England. They were valued as much for the perfume of their flowers as for their fruit. Early gardeners, knowing that citrus trees respond to stress by flowering out of season, deliberately stressed trees by not watering them for short periods in order to get at least two major flowering sequences with the accompanying perfume from each tree every year. Interest in citrus fruit intensified with the introduction of sweet oranges, limes, citrus monstrosities,[2] and their literature from Mediterranean countries.[3]

Orangeries and greenhouses

The ability to grow citrus trees at all in cool-wintered areas of the world was only available to the rich. Janson notes that in 1603, Henri 1V coined the term 'orangeries' to describe buildings that housed potted orange trees during the cold winter months.[4] 'Orangeries' became fashionable during the 17th and 18th centuries and were constructed in some of the famous gardens of Europe, including in the garden of Versailles in France built by Louis X1V.

The first buildings that could be described as orangeries, were built in Europe during the very late 1400s. They were simply closed warmed rooms with no extra light or ventilation used for sheltering potted citrus trees during the harsh winters. Over time, the design of orangeries developed to include windows or glazing of some kind on all sides (but not the roof area) to allow plenty of light. Some ingenious gardeners built orangery-type structures around in-ground plants each winter; these were dismantled when the weather again turned warm. Some later orangery designs were very similar to modern glasshouses with about 80% glazing.

During the 17th and 18th centuries, there were vast advances in greenhouse design and methods of heating. One driving force was the need for well-designed, unpolluted and safe orangeries. Greenhouses were also required for 'forcing' (see Glossary) plants to produce flowers and fruit out of season, as well as for the many new and exotic plants from different climatic zones in recently explored countries.

Orangeries of this period were generally heated by coal burning stoves. Potted plants were placed into wheel-mounted boxes or laboriously carried into and out of orangeries by hand. Some pots were so large that it took four people to lift and carry them. Elaborate pulleys or lifting devices were used to lift potted trees so that they could be knocked out and regularly repotted. Trees were watered with mixtures such as milk and honey to help revitalise them after repotting. This root drenching would have had the effect of making the plant produce many new fibrous roots quickly, similar to the effect of using liquid seaweed extract root drenches today.

CITRUS HISTORY IN NURSERY RHYME

A famous nursery rhyme and children's game song is 'Oranges and lemons' which evolved from the much more sinister and unpleasant 'Bells of London' rhyme.

'Oranges and lemons' say the Bells of St Clements
'You owe me five farthings' say the Bells of St Martins
'When will you pay me?' say the Bells of Old Bailey
'When I grow rich' say the Bells of Shoreditch
'When will that be?' say the Bells of Stepney
'I do not know' say the Great Bells of Bow.

There are fifteen bells in all recorded in the rhyme and their history is the history of 17th century London. The bells of St Clements are those of the small church of St Clements in London's Eastcheam near the wharves where citrus fruit was unloaded.

Spread

From the early 1500s, sweet oranges and the planting of orange groves spread from Spain to the Americas and, after colonisation, to Australia where some climatic conditions were favourable. Sometimes grafted plants or cuttings of citrus were shipped, but most often initial plantings in the USA and Australia used shipped seed to produce seedling trees and the different cultivars from these eventually became popular for home and commercial development.

Citrus growing in Australia

Australia was discovered and settled by Europeans at a time of intense exploration and interest in new species. Early settlers brought with them new and exotic species including citrus. They were unaware at the time that Australia had its own species of citrus. Groves of oranges using trees brought from Europe were established in Australia very early after settlement and fifty years on there was already a thriving citrus industry near Sydney. Sweet oranges were often sold as 'Parramatta' oranges because

CITRUS HISTORY IN SONG

The Australian bush song 'The Limejuice Tub' also known as 'The Whaler's Rhyme' refers to boats in its first verse as 'limejuice tubs', a direct reference to the use of limes as a scurvy preventive.

When shearing comes lay down your drums,
Step on the board you brand new drums,
With a rah-dum, dah-dum, rub-a-dub-dub,
We'll send you home in a lime-juice tub.

the Parramatta district was one of the first to be developed for citrus growing; some older Australians may well remember this name.

Citrus growing was found to be compatible with conditions in the drier, warmer areas of Australia, especially once suitable irrigation systems were designed towards the end of the 19th century. The expansion of the commercial fruit industry in Australia was quite rapid, and because of the favourable conditions and the high demand the citrus industry developed particularly rapidly. McAlpine[5] notes that 'for the 1896–97 period Victoria had only 32 acres under oranges and 107 under lemons outside of Mildura, which had 369 acres of the former and 445 of the latter, while in 1897–98 there were 43,261 orange and 58,522 lemon trees in the colony as against 27,835 of the former and 35,710 of the latter in Mildura.'[6] By 1910, 270,000 citrus trees were growing in Australia.[7]

Initial citrus tree plantings had disastrous results in many areas of Australia: many plants failed because of lack of appropriate management, and because of lack of understanding of new environments. Many early citrus plantings suffered badly or died out completely because of lack of understanding of Australian conditions resulting in poor site selection amongst other things. Suitable rootstocks were also lacking. Waterlogging, drought, frost, diseases and predation by pests dogged new plantings. Many trees developed root rots and collar rots especially in irrigated areas. McEwin describes the situation:

> In many cases the growers who were inexperienced irrigationists, ruined their trees with too heavy flooding of water, and the trees went 'sick' just at the time they reached the stage at which they should have been most profitable. Owing to want of sufficient knowledge as to the proper treatment of the [citrus] trees, many planters' blocks went so far back as to become worthless, and had to be uprooted.[8]

Some Australian insects such as the citrus gall wasp, and some animals (possums, parrots and kangaroos) quickly adapted to the introduced plant species and caused serious problems. Plant material brought to Australia in the early periods did not undergo any sort of quarantine, so many pests, diseases and viruses were introduced with plant material. None was more devastating than the phylloxera aphid plague that wiped out many areas of grape production after first being discovered at Geelong, Victoria, in 1875.

Nevertheless, the commercial citrus industry flourished so that by 1994 ten million trees had been planted (4.7 million Valencia and sweet oranges, 3 million Navel oranges, 1.2 million mandarins, 747,000 lemons, 244,000 grapefruit) producing half a million tonnes of fruit.[9]

Irrigation spurred the growth of the citrus industry. From 1887, the Chaffey brothers, originally from California, were prominent in developing huge irrigated citrus groves in the Murray River area from Mildura in Victoria to Renmark in South Australia. They were the first to grow 'Valencia' oranges in Australia and they also planted Navel orange trees. The Chaffey brothers utilised the waters of the Murray/Darling river system, using irrigation channels to carry water to orchards and vineyards to enable fruit production in dry areas with very low rainfall. Without this water supply, no horticultural enterprises could have been successful.

Salination of soils and other problems associated with flood irrigation were a continuing and increasing problem and many attempts were made to deal with these problems including installation of underground drainage systems, fan, drip and micro-irrigation methods. Irrigation technology has expanded to include water efficiency and savings through recycling.

CITRUS AND DEMENTIA

In a study reported by the BBC in August 2005, it was found that folic acid found in oranges, lemons and green vegetables could significantly reduce the risk of Alzheimer's disease. See www.alz.org for research report.

Continuing research

To improve commercial citrus production, State governments began horticultural and citrus tree research projects including those at the Horticultural Research Station at Irymple in northern Victoria, begun in 1954 and now named the Sunraysia Horticultural Research Station. A lot of the initial citrus rootstock research was carried out there as well as other citrus research at the CSIRO Division of Horticultural Research station at Merbein, in Victoria.

Beginning in 1955, a bud multiplication scheme was set up by the NSW Department of Agriculture at Dareton Horticultural Research Station to provide virus free buds for propagation purposes and to carry out research on the production of virus-resistant citrus rootstocks.[10]

Selections of citrus fruit occurring as natural mutations or 'sports' (see Glossary) and chance seedlings have been developed commercially since the citrus industry began. Some, such as the 'Leng' navel and Lane's

Irymple Research Station, July 1990

navel orange, are now important cultivars used within the industry today. The industry has also moved towards integrated pest and disease control systems (Integrated Pest Management or IPM) using biological controls and the application of far fewer chemical control products. In 1987, PVR (Plant Variety Rights) legislation was introduced allowing royalties for plant propagators, and later PBR (Plant Breeders' Rights) were introduced (see Glossary) allowing many different citrus cultivars to be introduced into Australia from overseas. Many new citrus cultivars will soon be seen in the market place, although many of the PBR registered cultivars have not been released for home garden use.

The citrus industry had hard times because of initial establishment costs and associated problems, low prices obtained for fruit or juice and competition from overseas products. Some growers during the 1970s bulldozed groves because they had become uneconomical. Even as late as the late 1980s and early 1990s, and again during 2005–06, cheap imported oranges from places such as California and cheap juice concentrates from Brazil, South Africa and Florida in particular flooded the market with a devastating effect on the industry. The industry fought back by providing 'fresh' oranges and 'fresh' orange juice marketed as 100% Australian. In 1993 the Australian Citrus Industry Council (ACIC) established a Code of Practice for the fruit juice industry to prevent adulteration and juice dilution and any unfair practices in the processing, packaging, labelling, marketing and reconstitution of fruit products.[11]

Citrus tree standards adopted by nurseries

1. Approved pest and disease management practice
2. Black plastic poly bag: diameter 15 cm, depth 30 cm
3. IPM approved potting medium
4. Well-grown root systems
5. Certified budwood
6. Disease-resistant rootstocks
7. Budding height range 30–35 cm
8. Minimum height 25 cm
9. Minimum height 60 cm
10. If nurserymen grow single-pole trees, they have the opportunity to top the tree to a set height (60 cm). In even light conditions, this will have the effect of forcing a uniform, strong crown.

(Source: Ian Tolley)

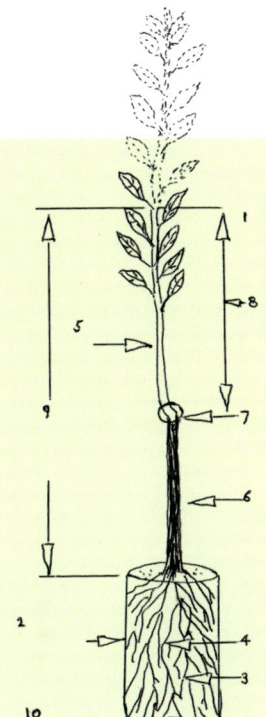

Citrus nurseries have adopted a set of standards to ensure that home gardeners and commercial growers can obtain the best quality trees possible.

Now due to research, better rootstocks, the availability of virus-tested rootstocks and virus free propagating material and recent technological advances citrus groves can be planted with a guarantee of success. Today, there are commercial citrus growing areas in all mainland States of Australia: in Darwin and Katherine in the Northern Territory; the Sunraysia and Mid-Murray regions of Victoria and NSW; Gosford, Narromine, Windsor, and Murray Irrigation Area of NSW, Carnarvon, Donnybrook, Gin Gin, Perth in Western Australia; the Riverland area of South Australia; and the Sunshine Coast, Gayndah/Mundubbera, Bundaberg, and Emerald in Queensland.

Although no commercial citrus production occurs in Tasmania, citrus trees perform well provided they are well-drained, in a sunny position sheltered from wind and frost. Most citrus tree species have evolved in tropical to semi tropical zones and often grow as under-story plants in forest areas. I have seen vast changes and increased vigour in the growth of a Meyer lemon grown in southern Tasmania with the erection of a shade-cloth mesh 'cage' around the tree to give wind protection. If the trees are given full protection of a controlled atmosphere greenhouse then the trees will grow as well as they do in tropical regions.

Australia's own citrus species including the 'Finger lime' have begun to be developed for commercial use.

Cumquat leaves, fruit and flower

Citrus products

Citrus products abound. Products range from the bitter 'Seville' oranges used for cooking and marmalade production to its cousin the 'Bergamot' orange used to provide the essential oil 'Neroli', an ingredient in the perfume Eau-de-Cologne. There are the sweet/sour grapefruit, large fruited citrons, pummelos and shaddocks used mainly for candied peel production. There are the citrus with religious significance such as the 'Etrog' and ornamental species such as the calamondin and cumquat for growing in pots. Very sour lemons and acidic but aromatic limes are available for flavouring drinks and food. The sweet-fruited citrus for eating fresh include mandarins, sweet oranges and hybrid fruits such as tangelos, together with blood oranges and grapefruit with blood-coloured flesh.

Commercial citrus products such as pure juice, frozen products, juice concentrate, lozenges, sweets, dried fruit slices, lemonade, icicles, ice-cream, alcoholic drinks, candied peel and glacé cumquats are readily available and there is a growing interest in citrus-based cleansers and allied products.

Pomander made from fruit and cloves

CHAPTER 2

citrus species and cultivars

It is not always easy to track down the origin of some citrus cultivars because of the thorough spread of citrus along historic trade routes, and the ability of many citrus species to cross-pollinate with other citrus species and throw 'sports' (mutate) to produce fruiting plants with observable differences to that of the parent.

The origins of the navel orange

The problem of tracking down the origin is highlighted by descriptions of the origins of the Washington navel orange cultivar. McEwin describes the navel orange as having been first introduced into the US in 1872 by William F. Judson, the United States Consul to Bahia in Brazil. Having learnt of orange trees with no seed growing in a swamp on a bank of the Amazon, some sixty miles inland, Judson arranged for some shoots to be collected, carefully packed in clay and moss, and forwarded to the Agricultural Department at Washington. Four survived, and were given to a Mr H. Tibbetts of Riverside, California, who had applied for any trees or shrubs for experimental growing in his district. Only two survived and began bearing in 1878. The few fruits they produced were very fine and large, and found to be seedless so Tibbetts propagated thousands of trees from them.[12]

Webber and Batchelor[13] describe the original navel orange tree material as having been collected from Bahia by another person completely: the Reverend F. I. C. Schneider, who first advised the Department of Agriculture about the orange's existence and potential in 1869. Later, a US Department of Agriculture representative[14] requested a few plants. According to Webber and Batchelor, Reverend Schneider sent shoots from this citrus from Brazil to the USA, but they dried out in transit.

Twelve budded trees were also shipped from Bahia to the USA and survived and, in 1873, Mrs L. C. Tibbets of Riverside, California, received two budded trees, with buds cut from the original imported trees and grew them in her garden. According to this version, the navel orange industry in the US developed from buds of Mrs L. C. Tibbets' trees and from there also spread to other countries.

These two very different accounts, with their confusion of names, people and dates, show just how difficult it is to work out the exact origins of the Washington navel. Two separate selections may have been sent at slightly different times to two different family members. There may have been confusion over the originally introduced 'Bahia' navel orange and the 'Washington' navel orange. It is recorded both in Australia and the USA that the 'Bahia' navel had been imported before the selection of the 'Washington' navel and the subsequent development of that industry. In fact, the 'Australian Navel' name was given to a group of navel orange types originating from the Brazilian 'Bahia' orange but developed in Australia.

> In China, during the Cultural Revolution, Chairman Mao Zedong's Red Guards cut down all citrus trees that they could find because they were considered 'bourgeois'. Fortunately, a few trees survived but some valuable species may have been lost forever.

Naming of cultivars

Some citrus cultivars have been given more than one name. Many citrus have the ability to cross-pollinate or throw 'sports' and bud mutations and these have sometimes been given the same name as the parent plant. As well, even selected buds used for propagation may change (mutate) from one generation to the next to give slightly differing fruits on the trees produced.

Several citrus species have had recent name changes, for example *Microcitrus* is now classified as *Citrus*. Plant Breeders' Rights and Plant Variety Rights (see PBR and PVR in Glossary) mean that cultivar names can now be registered for royalties.

Citrus industry standards

Australia has a coordinated national Citrus Improvement Program (ACIP) and budwood and rootstock seed schemes called Auscitrus,[15] the trading name of the Australian Citrus Propagation Association Incorporated (ACPAI).

Auscitrus is responsible for pre- and post-importation and release of high-health, true-to-type budwood and rootstock seed to the citrus industry. Auscitrus is an initiative of the Australian Citrus Propagation

Association (ACPA), the Australian Citrus Improvement Association (ACIA), the Horticultural Research and Development Corporation (HRDC), and NSW Agriculture. Auscitrus is responsible for:
- Budwood and rootstock seed production and distribution.[16]
- Pathogen indexing for viruses and diseases and their elimination.
- Maintenance of virus free and immunised foundation trees in insect proof screen houses.
- Importation of new citrus cultivars.
- The horticultural screening (field testing) for all new citrus cultivars (see also Glossary).

Note that some cultivar material available from Auscitrus is not completely pathogen free because no pathogen free material of that cultivar exists in Australia. Note also that Tristeza virus is endemic in Australian citrus, being readily transmitted by the black citrus aphid.

The various species of citrus and their cultivars

The citrus genus of plants has about twenty species and this includes plants formally identified as *Erimocitrus*, *Fortunella* and *Microcitrus*.

Botanically, citrus plants are identified and classified by examining flowers, fruit and foliage of the mature trees. Generally, citrus leaves are dark green and shiny and have prominent oil glands. Leaves can be trifoliate (three leaflets per leaf) as with the related species Trifoliate orange (*Poncirus trifoliata*), which has ancestral connections to citrus species. Trifoliate orange is classified as a single species and will form viable seed when cross-pollinated with citrus. Trifoliate orange rootstock is used for various citrus.

Careful examination of the leaf of many species will show two of the three leaflets are reduced to narrow green sections or blade shapes (wings) that occur on the lower segment of the double-segmented leaf petiole (stalk). Some citrus species have two seemingly jointed leaves as shown on the Kaffir lime (*Citrus hystrix*). Others, such as lemons, have one leaf with winged or wingless leaf stalks.

Citrus flowers are fragrant and star-shaped, waxy in appearance and usually occur singularly but can occur in clusters.

Citrus fruit or hesperidium (the botanical term) is actually a form of berry, and can be oval, rounded or have a flattened-round shape. The skin is spongy, has oil glands and varies in thickness from a few millimetres (as on some mandarins) to several centimetres (as on pummelos). With some mandarin cultivars, the skin is very easily removed and can seem detached from the inside flesh (pulp). Underneath citrus fruit skin is a layer of a white pithy substance. At the fruit centre are cellular, banana shaped segments arranged in a rough sphere. These segments are often enclosed in web-like pith.

Citrus tree species and their fruit vary tremendously throughout the world. Some produce almost inedible sour or dry non-juicy fruit, while others produce fruit that can vary in size, taste and colour. Tree size may also vary: from shrubby types to large trees. Tree variation is often dependent on the soil the plant is growing in, the rootstock used, climate, water supply and level of available nutrients. Leaf shapes and sizes also vary. Flowering and ripening period of cultivars may differ in different climates and many species have several cultivars that ripen over an extended period of time.

Citrus categories according to species are as follows:
- *Citrus aurantiifolia* syn *Limonia aurantiifolia*, Lime, a tree 4.5m tall and 3m wide.[17]
- *Citrus aurantium* syn *Citrus × paradisi*, *Citrus sinensis*, *Citrus × tangelo*, *Citrus × tangor*, which includes the grapefruit group, sour orange group, sweet orange, Tangelo, Tangor and Chinotto. These trees can grow to 4–10m with a spread of 4m.
- *Citrus australasica* (Finger lime), *Citrus australis* (Wild lime, Round lime), *Citrus garrawayae* (Mount White lime), *Citrus inodora* (Russell River lime), *Citrus glauca* syn *Erimocitrus glauca* (Desert lime). Small trees up to 10m tall and 4m wide.
- *Citrus hystrix*, Kaffir lime, Leech lime, Makrut lime, up to 3m tall and 2m wide.
- *Citrus jambhiri*, Rough lemon ('Citronelle'), 5m tall and 4m wide.
- *Citrus japonica* syn *Fortunella japonica*, *F. marginata*, Cumquat up to 3m tall and 3m wide.
- *Citrus × latifolia*, Tahitian lime, 3m tall and 3m wide.
- *Citrus limetta*, Sweet lime, 2.5 tall and 3m wide.
- *Citrus limon*, Lemon, 5m tall and 4m wide.
- *Citrus limonia*, Rangpur lime, 4m tall and 3m wide.
- *Citrus maxima*, Pummelo, 12m tall and 4m wide.
- *Citrus medica*, Citron, 5m tall and 3m wide.
- *Citrus × meyeri*, Meyer lemon, 4m tall and 4m wide.
- *Citrus × microcarpa* syn *× Citrofortunella microcarpa*, Calamondin, Panama orange, 8m tall and 4m wide.
- *Citrus reticulata*, Mandarin, and hybrids Satsuma, Tangerine, Clementine 5m tall and 4m wide.

I have grouped the various citrus somewhat differently with more of an emphasis on taste and use as this is generally the way home gardeners approach fruit growing. For ease of identifying sweet, sour and different shaped citrus fruits I have grouped the citrus dealt with in this book into five main groups: sweet oranges and other sweet citrus; lemons and other sour citrus; grapefruit and large citrus; citrus grown for their leaves; and Australian citrus species.

Sweet oranges and other sweet citrus

Oranges (*Citrus sinensis*) are the sweetest of the citrus and probably the most popular as well as the most significant commercially. This group of citrus contains many species including interbred hybrids such as tangelos. In the earliest days of commercial citrus plantings, sweet citrus plants were used as seedling rootstocks for other citrus species. Trees vary in shape, size, and fruit production. Sweet orange trees can grow to 10m or more, have glossy, shiny leaves and produce white, strongly scented flowers.

Various citrus species are listed under the sweet orange category. For the purposes of this book I have grouped oranges with other sweet citrus including mandarins (*Citrus reticulata*), tangelos (*Citrus aurantium* syn *Citrus reticulata* × *Citrus paradisi* syn *Citrus* × *tangelo*), and tangors (*Citrus aurantium* syn *Citrus reticulata* × *Citrus sinensis* syn *Citrus* × *tangor*).

Most sweet oranges known today have their origin in China. Some were spread as seed along the old trading routes while much more recently others, for example the 'Emperor of Canton' (Emperor) mandarin and the 'Meyer' lemon have been imported direct from China. Citrus hybrids such as tangelos and tangors have been especially bred in citrus breeding programs in the USA.

Sweet oranges (*Citrus sinensis* **cultivars**) Early Australian orange orchards were developed from seedling planted trees and because there was a lot of variation in the fruit produced by seedling trees, different cultivars have evolved over time. Sweet oranges originally imported into Australia for commercial use included 'Jaffa', 'Joppa', 'Siletta', 'St Michael', 'Parramatta' ('Mediterranean Sweet'). As a general rule, these sweet orange cultivars produce fruit of medium size (an exception is 'Jaffa' which produces medium to large fruit), round or slightly oval with thin skins, very juicy with a very sweet flavour. Even though these trees generally crop very well, they are not now commercially grown in Australia to any great extent, probably because of the size of their fruit, their many seeds and the fact that other, later introduced oranges such as 'Valencia' (from 1870s onward) are better for juicing and storage. Most of these older cultivars are, however, still available to Australian home gardeners. Also available are cultivars of 'Valencia' oranges, and a range of early and late navel oranges. American seedling or 'sport' selections of sweet oranges include 'Parson Brown', 'Pineapple' and 'Hamlin', and are also available to growers and home gardeners in Australia. These American selections

Mediterranean Sweet orange fruit

have, in the past, been judged as amongst the best sweet oranges in the world.

Commercial growers now concentrate mainly on 'Valencia' orange cultivars, navel oranges and selected mandarin cultivars (see below).

Navel oranges (*Citrus sinensis* cultivars) The navel orange was imported from Brazil in the early 1870s by the US Department of Agriculture in Washington under the name 'Bahia', and was eventually distributed to growers from there, thus resulting in the now common name 'Washington Navel' orange. The 'Washington Navel' was imported into Australia in the late 1880s.

Navel oranges are the sweetest of all sweet oranges. They are usually seedless with easily peeled skins and a characteristic 'navel' shaped lump at the base of the fruit. This navel often protrudes into the fruit and has miniature slices of orange pieces grouped into a pyramid inside the fruit; these are edible and do not detract from the overall appeal of the fruit. Breeding has eliminated this 'Navel' characteristic, undesirable because the navel often harboured mealybug insects.

Navel oranges have been one of the most commercially successful of the sweet oranges grown in Australia (see the table of 'Navel orange cultivars', below) and this has contributed to the search for and development of new cultivars. These have been appearing for many years with more local selections of superior cultivars having been made in recent years. Some of the cultivars listed in the table below have derived from sports or mutations of such cultivars as 'Washington' navels. Increasingly, new cultivars are subject to Plant Variety Rights – PVR, and Plant Breeders' Rights – PBR (see Glossary) making many of the new cultivars unavailable to the home gardener. There are other cultivars still being developed and trialled.

Navel orange cultivars (many not available to home gardeners)

Atwood	This cultivar was imported from California in 1990 where it is an old budline originating as a sport (natural mutation) of the Washington navel. It is an early to mid-season navel, large and round, and very orange in colour.
Autumn Gold	This cultivar was identified in the early 1950s by Jack Pollock, from Strathmerton in Victoria. It is a seedless, late maturing navel, smaller in size than others such as Lane's Late.
Barnfield	This cultivar originated in the early 1980s on the property of Wayne Barnfield in Ellerslie, NSW. It is a seedless, late maturing navel with good flavour.
Chislett	This navel originated on the Chislett property at Kenley in Victoria. It is a seedless, late maturing orange with good tree-holding qualities.
Fukumoto	This is another navel introduced from California where it is grown widely as an early- to mid-season fruiting cultivar. It was released from plant quarantine only in 1999 so it is a relatively late addition to navel cultivars in Australia. It's fruit is medium sized and round with a relatively small navel.
Koala™ Easy Peel	This is a relatively new trademarked cultivar from Belevedere Fruit Growers in north-western Victoria. It has sweet, seedless fruit low in acid, easily removable skin, and is easy to separate into segments.
Lane's Late	This navel has very large fruit. It originated on the Lane property at Curlwaa in NSW from a limb mutation of Washington navel. It is another late maturing, seedless navel with some distinct differences from the original Washington, with a thinner and smoother rind and a less obvious navel.

Navel orange cultivars (continued)

Leng	This cultivar originates from an early ripening sport found growing on the Leng property at Irymple Victoria in 1934. It has thinner skin and is very suitable for juicing.
Navelate	Navelate is a cultivar imported into Australia in the 1980s, and originating in Spain in 1948 as a mutation of the Washington navel. It is a seedless mid to late season navel with smallish fruit and concealed navels.
Navelina 7.5	This is another cultivar imported from Spain in the 1980s. It originated in Spain in 1910 and is widely grown there. It is an early maturing, seedless navel with large fruit of somewhat variable shape and of a deep orange colour.
Summer Gold	The cultivar was originally developed from a selection found by Dudley Marrows in the 1950s at Mourquong in NSW. It is a mid- to late-season seedless navel with a consistent fruit size making it a good commercial option because it prolongs the picking season.

Foliage and fruit of 'Navel' orange

Valencia oranges (*Citrus sinensis* 'Valencia')
The 'Valencia' orange introduced into Australia was first grown in the USA in the 1870s and was initially known as 'Rivers Late' after the plant nursery that supplied the unnamed plant from England. It was later identified as the 'Naranja tarde de Valencia' orange from Spain and then given the name 'Late Valencia' because of its ripening characteristics.

It was one of the first commercial citrus fruits introduced, and the Chaffey brothers were the first to grow 'Valencia' orange trees commercially. The Chaffey brothers established irrigated orange groves in Renmark, South Australia and Mildura, Victoria in about 1887. Until recently, Valencia was still the orange with the highest commercial production in Australia, but now navel oranges are catching up.

'Valencia Orange' showing effects of regreening of skin colour

Several commercial selections of late and early ripening cultivars of this orange have been developed in Australia since, to extend the picking season and supply of fruit throughout the year. The Valencia is a medium-sized orange with a thin skin and slightly acidic juice. It is an acceptable fruit fresh from the tree. It keeps well and produces excellent juice used in the commercial fruit juice trade. Some good commercial cultivars are 'Keenan's Valencia', 'Valencia Seedless' and 'Late Valencia'.

Mandarins, Tangelos and Tangors This is a mixed group of sweet citrus that consists of true mandarins (*Citrus reticulata*) and hybrids of several species and cultivars including Tangelos 'Orlando', 'Minneola', 'Seminole' (*Citrus aurantium* syn *Citrus reticulata* × *Citrus paradisi* syn *C.* × *tangelo*) and Tangors (*Citrus aurantium* syn *Citrus reticulata* × *Citrus sinensis* syn *C.* × *tangor*). The taxonomy in this group is confusing as many of the named cultivars have been bred from cultivars of mandarins some of which have been crossed with other species and so, as with other citrus, it is sometimes difficult to trace the genetic parentage. Hybrids within this group have been given names such as Clementine, Satsuma and Tangerine.

Native to China, true mandarins (*Citrus reticulata* syn *Citrus nobilis*) seem to have been the major citrus grown for centuries in India, China

BLOOD ORANGES AND OTHER PINK- OR RED-FLESHED CITRUS

These are selections or 'sports' that have appeared from seedlings, natural crosses or mutations. Most come from the Mediterranean region. They have pink or red pigmented flesh that develops best when trees are subject to hot dry climates with cool nights. Some of these cultivars have a red-coloured blush or stripes on the fruit skins. In general, they are usually sweeter and may be juicier than other cultivars of the same type.

Sliced fruit of 'Blood Orange' and 'Red Ruby' grapefruit

and Japan. I have seen many different cultivars in China including fruit with emerald green skins when ripe, but this is a normal occurrence as there is always little colour on fruit grown in tropical to sub-tropical regions. Some cultivars produce more fruit and seed when pollinated with another cultivar.

The trees in this group of citrus can reach 5 metres with a rounded shape and are very hardy and cold-tolerant. Leaves are small, with a pointed blade shape, and finely serrated.

Fruit is round or flattened-round, 3–10cm wide, usually with smooth and shiny, orange-red, loose skin. Fruit segments are easily pulled apart although some cultivars have 'tight' skins.

Several well-established mandarin, Tangelo and Tangor cultivars are grown in Australia, a selection of which is included in the table below. There are other new introductions becoming commercially available, some of which are also included in the following table.

Mandarin cultivars (asterisked when available to home gardeners)

Afourer
(syn 'W. Murcott', 'Nadorcott', 'Delite')*
This is a late ripening Tangor, the original selection of which was made in Morocco. It has only very recently been released from quarantine in Australia so there are few commercial plantings as yet and it is not widely available. The trees are vigorous and produce low seeded fruit if isolated from other species. The trees fruit within five years of grafting/budding, producing medium-sized fruit with bright orange flesh and good flavour.

Avana Apireno
This is a mid-season ripening mandarin, one of the commercial cultivars grown in Corsica, Italy. The fruit produced is similar to the Imperial mandarin and matures at the same time, it has lower seed count per fruit and better fruit and skin qualities than the Imperial mandarin. This is a registered commercial cultivar.

Avana Tardivo di Ciaculli
(syn 'Avana Tardivo')
This is one of the late-ripening mandarin cultivars that is commonly grown commercially in Italy. The fruit has fairly high seed content (15–20 per fruit) and a high juice ratio per fruit. This is a registered commercial cultivar.

Clementines
These are classified as a type of mandarin. They are specifically a group of different Mediterranean mandarin cultivars that have been especially bred and improved to be more acceptable to consumers. Traits such as increased fruit size, brighter skin colour and seedlessness have been achieved. Although almost seedless, the fruit tend to produce more seed if not isolated from other citrus cultivars. Clementines are subject to alternate cropping; pruning after a light crop will help prevent this trend. The preferred rootstock is Trifoliata.

Ellendale
(or Ellendale Beauty)*
This Tangor was originally found growing as a seedling on the Burgess property in Burrum, Queensland, in 1878. It is a late season, seedy variety that, for some time, was the chief export variety of mandarin. It is possible that it is a natural mandarin/orange hybrid. The fruit is large, slightly flattened, bright orange in colour, easily skinned and very juicy with an orange-mandarin flavour.

Ellenor*
This is an Ellendale hybrid developed in the 1940s. The fruit is fairly small and has a coarse texture, although it is very juicy and tasty.

Emperor
(syn 'Emperor of Canton')*
Thought to have been brought from China during the gold rush period, this is a mid-season cultivar with yellow-orange, often dimpled loose, oily skin and pleasant flavoured pulp.

Mandarin cultivars (continued)

Fallglo
This cultivar is a complex hybrid of mandarin, orange and grapefruit, originally bred in Florida in the US. It is an early-maturing, highly seeded variety.

Glen Retreat*
(syn 'Beauty')
The parent tree was grown in the orchard of a Mr W.H. Parker of Glen Retreat near Brisbane about 1921 and has become one of the most popular mid-season mandarins. It is large and bright orange, very seedy and with a good flavour. The tree is a heavy bearer but an alternate cropper.

Hickson*
This cultivar was the result of a limb 'sport' on an 'Ellendale' tree growing on the Hickson property in Roma, Queensland. It is widely grown as a commercial variety in Queensland. It is similar to Ellendale but it has a small nipple like growth at the stem end, a very smooth, glossy yellow-orange coloured skin. The fruit has few seeds and peels very readily.

Imperial
(syn Early Imperial)*
This is a very popular commercial and home garden cultivar, originating from a chance hybrid at Emu Plains in NSW. It is an early mandarin, seeded, with smooth easily peeled skin. The flavour is good and the segments separate easily. The fruit can be picked over many weeks.

Kara*
A popular cultivar that fruits in early summer and produces very bright-coloured orange, easy to peel fruit that has few seeds and a great flavour. The fruit will hold on the tree for an extended period compared with most mandarin cultivars.

Karra Organic Farm mandarins

Minneola*
This is a Tangelo with large bright red-orange fruit with a glossy thin skin, ripening mid-season. Unfortunately, the skin is difficult to remove and sometimes tears flesh away but the taste is great for those that like the grapefruit flavour. The tree tends to bear only every second year (biennial bearing).

Mandarin cultivars (continued)

Mor (syn 'Seedless Murcott')	This is a mandarin originating in Israel. The fruit has very few seeds, an excellent sweet taste. This is a registered commercial cultivar.
Murcott*	This is a Tangor originating as a mandarin/sweet orange hybrid in the USA and originally introduced into Australia illegally as budwood in the 1970s. Since then it has become one of the most popular cultivars for export and domestically. It is a highly seeded, large-fruited cultivar, maturing late season, juicy and with good flavour.

Or (mandarin hybrid)	Originating in Israel, this tree produces very sweet fruit with easily removed skin that is of good size with highly coloured flesh and stores well. This is a registered commercial cultivar.
Page*	This is a 'Minneola' tangelo and Clementine mandarin hybrid originally from the USA. It is an early- to mid-season fruit, very seedy, that looks a bit like an orange. It is popular for home gardens although it is self-sterile and needs a pollinator. The flavour is excellent and fruit very juicy.
Satsuma*	This popular cultivar produces smooth-skinned very flavoursome, seedless fruit that is easily peeled. Fruiting is in May and the tree is suitable for cooler climates.
Seminole*	This Tangelo was bred in the USA; it ripens late and has similar fruit characteristics to 'Minneola' except that the skin is bright orange and the tree tends to over-crop producing small fruit. The home gardener growing this cultivar will find that thinning of the fruit is necessary.
Tacle	This is a mandarin hybrid originating in Israel. The fruit is easily peeled with very smooth, dark orange skins and a red flesh colour.
Winola	This is another mandarin hybrid originating in Israel. The medium-sized fruit has a deep reddish-orange coloured skin.

Lemons and other sour citrus fruit

The lemon (*Citrus limon*) is probably the fruit that most people associate with sour tasting citrus, although there are degrees of sourness even amongst different cultivars of lemons, with some cultivars being quite sweet. For the purposes of this book, I have grouped lemons with other sour citrus including the Rough lemon (*Citrus jambhiri*) which is not a true lemon; sour oranges (*Citrus aurantium*); the citron (*Citrus medica*); the calamondin or musk lime (*Citrus microcarpa*); chinotto (*Citrus aurantium* syn *Citrus myrtifolia*); cumquats (*Citrus japonica*); limes (*Citrus aurantiifolia*); and Kaffir limes (*Citrus hystrix*).

Palestine Citron

Lemons (*Citrus limon*) As with other citrus, the exact origin of the lemon is not easily traced, but it is thought that it originally came from the warm areas of the Himalayas, Burma and Southern China and was taken to Palestine in the thirteenth century before being brought to Europe. Some botanists think, however, that the lemon is a hybrid plant that evolved in the Mediterranean area. Columbus, on a voyage in 1493, was responsible for taking citrus including seeds of lemons to the West Indies where some were planted and thrived. Then the growing of citrus including lemons spread still further from there to other parts of the world including Australia.

Of all the citrus, the lemon probably has the most uses but it is not grown as much commercially as some other sweet fruited citrus such as 'Valencia' oranges which are used for the vast orange juice and juice concentrate trade.

Lemon trees are usually smallish in size. On some cultivars the tree branches and stems have annoying spiky thorns. Leaves vary in size and shape: they can be small or large, winged or wingless leaves but are generally aromatic, pale green and can be pointed with smoothly serrated leaf margins. Flowers are perfumed and pinkish on the outside. Fruit is yellow, sour to acidic, oval, and can vary in length from 7 to 15cm. Fruit generally has a terminal nipple that can be prominent or suppressed.

Lemons will grow in every State of Australia with regular watering in summer, so just about every home gardener with sufficient space can grow a lemon tree somewhere in the garden or even in a pot.

Many different cultivars have been developed or chosen from seedlings, mutations or sports, both here and in other countries since the lemon was introduced. William Watson in *The Gardener's Assistant*[18] lists lemon varieties available in 1936 as being 'Bijou', 'Imperial', 'Common', 'Sweet', 'Metford's' and 'White'. Most of these cultivars are not available today, as newer selections have replaced them. Some of the cultivars now most commonly grown by home gardeners in Australia are shown in the following table.

Lemon cultivar	Description
Eureka (*Citrus* × *limon* 'Eureka')	The Eureka lemon originated as a chance seedling in the USA and was selected for its smooth-skinned fruit, their juiciness and ability to store well. This lemon tree has fruit with a smooth feeling but lumpy skin compared to cultivars such as 'Meyer' although the fruit is usually larger in size and has few seeds. The lemons have a high juice content, taste sharply acidic and the fruit matures almost all year round. One of the most distinctive features is that the trees have no thorns.
Lisbon (*Citrus limon* 'Lisbon')	This tree seems to have occurred originally in Australia about the mid-1800s and was thought to have been introduced to Australia by the Portuguese. 'Lisbon' lemon fruits are smoother than those of 'Eureka', and are produced mainly during the winter period with a smaller crop over the summer. The tree has extremely sharp thorns, but it seems to be more cold-tolerant than others. Thornless cultivars could soon be available.

Variegated Lemon (*Citrus limon* 'Variegated')	This lemon has yellow and light green variegated leaves and is mostly used for ornamental purposes. It does produce decorative, yellow-skinned lemons with light green stripes along the fruit and these can be used as ordinary lemons for their juice. According to some, this fruit has the best flavour of any lemon available.

Lemon cultivar (continued)

Villa Franca
(*Citrus limon* 'Villa Franca')

This lemon originated in Europe and is one of the favoured lemons for tropical regions. The fruit can be harvested year round and is very acidic and similar in shape and size to 'Lisbon' lemons, although the tree has fewer thorns than 'Lisbon'. It is favoured for its summer cropping.

Villa Franca lemon (silvering of fruit typical of mite or rubbing leaf damage)

Lemonade
(*Citrus limon* 'Lemonade')

This is thought to be a hybrid between the Meyer lemon and an orange. The fruit has a taste that is pleasantly sweetish and not very sour or acidic. The large round yellow fruit is thin-skinned and can be eaten as you would an orange. Juice from fruit makes a great 'lemonade' without the acidic 'zing' of truly sour lemon juice. The Lemonade tree grows well in tropical climates as well as in cooler regions. It will fruit from an early age and produces fruit through the autumn and winter. The mature tree is relatively small so it is suitable for small garden areas and pots.

Meyer
(*Citrus* × *meyeri* 'Meyer')

This lemon was named after the American plant collector Frans N. Meyer (1875–1918), who introduced it from China. The fruit is mildly acidic and can be eaten without the sour taste of most lemons. The skin is very smooth and can become a bright lemony-orange colour giving the impression that the tree may be the result of a natural cross between a lemon and an orange tree. The 'Meyer' is an excellent tree to grow in pots as it tends to be smaller growing than other lemons. The tree is very frost hardy and tolerant of cold conditions and has some ripe fruit upon the tree most of the year.

LEMON USES

- Copper can be cleaned with a mixture of lemon juice and salt.
- Scratches on wood can be removed with equal parts lemon juice and salad oil rubbed in.
- Lemon juice gives shine and sparkle to glass windows.
- To remove discolouration from inside aluminium cookware boil water and lemon juice together.
- Cut lemons can be used as a deodoriser in refrigerators and rubbish bins.
- Stop eggs from cracking when boiled by putting some lemon segments in the water with the eggs.
- Clean and wash out refrigerator using lemon juice and water.
- To keep cauliflowers white add strips of lemon rind and juice to the cooking water.
- Make sour cream by adding one teaspoon of lemon juice to each two cups of cream.
- Remove scorch marks from some fabrics (especially linen) by rubbing the stained area with lemon juice and allowing it to dry in bright sunshine.
- Mildew can be removed from washable fabric by applying lemon juice to the mildew and allowing it to dry in the sun.
- Inks and some stains can be removed from hands/fingers by applying lemon juice.
- Lemon essence can remove ballpoint marks on synthetic materials.
- Crayon marks on non-washable wallpaper can be removed using lemon essence.
- Rust stains on washable material can be removed with a mixture of lemon juice and baking soda.
- Microwaving whole lemons for a few seconds will increase the amount of juice extracted.
- Remove fish smells from hands after a days fishing with lemon juice.
- Home-grown lemons will keep longer if they are removed from the tree with a short stalk attached.
- Rub lemon juice onto cut fruits such as apples and avocado to stop them oxidising and becoming brown.
- Freeze lemon juice in ice cube trays for later use.

Citron (*Citrus medica*) 'Medica' in the species name does not indicate that this citrus has medicinal properties. Citrons were probably the first citrus species taken out of South-East Asia to the Middle East and Europe centuries ago.

Citron trees vary in size according to the cultivar but some grow into largish trees. They have a straggly growth habit and generally have thorns. Trees are frost and temperature sensitive, suitable for a limited climate range. Leaves are single oval with wingless leaf stalks. Flower petals have a purple tinge on the outside surface. Citrons are a citrus species readily grown from cuttings.

Fruit is usually about twice the size of an average lemon, although there are various small- and large-fruited cultivars as well as unusual shaped cultivars throughout the world. These include the 'Etrog' and 'Buddha's Hand' (Fingered Form). Fruit is yellowish to bright yellow at maturity. Some have smooth skins, others rough skins and they have a thick, fragrant skin and very little flesh (pulp), what little there is often being dry and usually very acidic. As peel or rind makes up most of the fruit volume of citrons, most cultivars are grown specifically for production of dried and candied peel. Citron are not grown to any great extent in Australia.

There are sweeter cultivars such as the 'Corsican' citron: it produces large (diameter 7.5–10cm and length 10–17.5cm) and elliptical fruit, with rind about 2cm thick.

The *Citrus medica* 'Buddha's Hand' (and 'Fingered form') originated in India although it is grown in other places. It looks like and is about the same size as a human hand with many finger-like growths extending from the fruit. In Australia, it is mainly grown as an ornamental tree. The skin of the fruit is highly aromatic.

'Etrog' (or 'Esrog') is a citron with an interesting history. Its origins are unknown but it is recorded as far back as 4000 BC and is common in countries around the Mediterranean. The 'Etrog' is a relatively small citron, with fruit a little larger than an average lemon. The fruit is oval with a gradually sloping, pointed end opposite the stem end. Fruit often has a persistent stylar growth (called a pitam) at this end. This stylar growth is seen on many citrus species, but it usually withers away or falls off before the fruit matures. With many Esrog fruit the pitam remains.

The Etrog citron has a role in Jewish Feasts of the Tabernacles. Each participant takes selections of four species of plant: palm leaves, twigs of myrtle, willow shoots and 'Etrog' citron to these festivities. Citron must be kosher; it must be of perfect shape with a prominent pitam, the skin must be only of certain colours and not blemished. Green coloured fruit must have part of the skin turning yellow, fruit cannot be taken from a grafted tree. Bruised fruits are not acceptable, crooked bent or deformed fruits are not used. This list continues.

Citron fruit

Citrus 'Buddha's Hand'

Rough lemon (Citronelle)

Rough lemon (*'Citronelle'*) (*Citrus jambhiri*) Botanically, Rough lemons are not described as true lemons, but they are sometimes used as lemons or in jams such as marmalade. The tree may be a hybrid between a lemon and a citron.

Rough lemons are thought to be amongst the first citrus to have been spread by seed to nearly all parts of the world. In some areas, for example in parts of the USA, it has become naturalised and can be found growing as a seemingly 'natural' forest tree.

This tree is tough and grows fairly 'true to type' from seed. It was used as a rootstock for other citrus (see section on rootstocks Chapter 3). Its small fruit is very lumpy, with thick skins and flesh that does not contain much juice. A tree laden with fruit makes a spectacular sight, making the tree an excellent ornamental.

The Rough lemon is susceptible to the citrus gall wasp and this can be a problem (see Chapter 6 on pests and diseases).

Seville oranges (*Citrus aurantium*) *Citrus aurantium* is known as 'Seville orange' and 'Bigarade' as well as 'Sour orange' or 'Bitter orange', and is one of the first citrus species grown commercially in Europe outside its place of origin in South-East Asia. It was planted extensively around Seville in

Smooth-skinned Seville orange

Spain, before sweet oranges replaced sour oranges as the citrus of choice. Sour oranges were, and still are, used for cooking as they are usually too bitter to eat fresh, although there are sweeter and milder sour orange cultivars.

Sour orange trees make very attractive specimens. Trees are tough and fairly frost-resistant, and can grow to about 10 metres. The foliage is dense and aromatic and leaves are sharply pointed, 7–10cm long, and winged. There are spines on the limbs. The flowers are very white and scented. Fruit are globular, slightly flattened, and 7–8cm diameter or larger. Some sour orange cultivars have rough skins, others smooth. 'Rough Seville' trees fruit in late winter and produce medium-sized oblong to round shaped yellow fruit with a rough bubbly skin. 'Smooth Red Seville' also fruits in the late winter period; the fruit is highly coloured, smooth-skinned and medium to large in size and of a round but flattened shape.

The pulp of the fruit is acidic to sour and the membrane very bitter in some cultivars. The segmented fruit core is hollow.

When Seville oranges are boiled, their skins become transparent and the pulp, when cooked with sugar, sets readily into a jelly, making them ideal for marmalade, one of their main uses. Their peel is used for production of liqueurs such as Grand Marnier, Cointreau and Curacao. The dried peel is used medicinally, and oil from the skins is used in perfumery. Leaves and flowers of sour oranges also yield oil used in perfumery and as a flavouring agent.

The sour orange has been used as a rootstock for other citrus trees (see section on rootstocks in Chapter 3, and Glossary) and is thought to be a parent of Poorman orange (*Citrus × aurantium* cultivars), a largish grapefruit type fruit with thin orange-yellow skin that ripens earlier than most grapefruit and is a hardy plant especially when grown on trifoliata (*Poncirus trifoliata*) rootstock. The Poorman orange has orange-coloured flesh and a pleasant flavour without too much bitterness and is generally sold as a grapefruit especially in New Zealand. The Poorman cultivar was brought to Australia in about 1850 from China and then taken to New Zealand, where seedless cultivars were developed. The Poorman fruit selections are sold in New Zealand and Australia under the names of 'Poorman', 'Kawau', 'New Zealand Grapefruit', 'Morrison's' and 'Goldfruit'.

Calamondin fruit and foliage

The sour group of oranges also includes the Bergamot orange (*Citrus aurantium* ssp. 'Bergamia'), a selection of the sour orange growing to about 5m. This tree has been grown in the Mediterranean area since the seventeenth century, and is famous for the oil produced from the fruit peel. Oil of Bergamot is used primarily in perfumery particularly in Eau de Cologne. The oil is mainly produced in Southern France, Italy and Paraguay.

The Bergamot orange grows best in semi-tropical regions, although it will grow well in cooler areas if given protection. The Bergamot orange tree has small spherical fruit with smooth, bright yellow skin; they yield a sour and not very palatable juice but the skins and flesh can be used for marmalade.

Calamondin (*Citrus microcarpa* 'Calamondin', syn *Citrus mitis* syn *Citrus madurensis* syn *Citrofortunella microcarpa*)

Also known as Musk lime (and as the Panama orange, Scarlet lime and Golden lime), this tree, originating from China, has very small fruit not unlike the round cumquat (*Citrus japonica*). The fruit has a loose, bright orange skin, much like the mandarin, with seed that has many nuclei (called polyembryonic seed; see Glossary). The fruit pulp is sour and not very palatable.

The calamondin can grow to a large tree (to 8 metres); it is upright, cold-tolerant with small, tightly bunched foliage. The pointed leaves are smaller than most citrus. Because of its bright fruit it is mostly grown as an ornamental, although the fruit can be used for preserving or bottling. A small growing, variegated ornamental selection of this tree with paler fruit is very popular for growing in pots.

Chinotto (*Citrus aurantium* syn *Citrus myrtifolia*) A tree commonly used as a pot species, the Chinotto has sharply pointed, smallish leaves in dense clusters. The small fruit is orange with bubbly skin.

Common names of this plant vary; it has also been called the Myrtle-leaf orange because the leaves resemble myrtle-tree foliage. The leaves are clustered together because the internodes are short, which gives the plant a bushy effect and explains why it is an attractive plant to grow in pots in small gardens or on patios. The slightly flattened fruit resemble those of a true mandarin fruit, so the plant has also been called a mandarin. The fruit is juicy but sour and is not usually eaten but will hang on the tree for an extended period providing a colourful show. The skin has been used to make candied fruit and the whole fruit can be used for marmalade.

Cumquats (*Citrus japonica* syn *Fortunella* spp.) Cumquats are believed to originate from China, although they have not been found growing in the wild. The plant name 'Fortunella' was given in honour of Robert Fortune, the Scottish plant collector,[19] who introduced the cumquat to Europe. The cumquat had already been cultivated for a very long time in China, Japan, and other countries of the Asia–Pacific area. There are many cultivars of cumquat around the world.

Close up of 'Nagami' cumquat

The cumquat (or Kumquat) tree is small (to about 3 metres) and very cold-tolerant. The leaves are clustered, small to medium size and slightly

Cumquat growing in tub ('Marumi')

rounded at the apex, tapering both ends with short leaf stalks. Fruits are yellow-orange, small, round or oblong and 3–7cm in diameter. Oil glands in the skin give the slightly sweet skin an oily taste. The pulp is slightly to very acidic.

The 'Round' or 'Marumi' cumquat (*Citrus japonica* 'Marumi', syn *F. japonica*) has small oval, slightly flattened fruit with small and pointed leaves. The fruit is used specifically for marmalade and for cumquat liquor or preserved glacé cumquats. The 'Oval' or 'Nagami' cumquat (*Citrus japonica* 'Nagami', syn *F. margarita*) has oval, highly coloured orange fruit with a thin skin. Both cultivars produce fruit that is very juicy and only slightly acidic.

Both cumquat cultivars can be eaten whole without removing the sweet skin.

'Tahitian' lime fruit

Limes (*Citrus aurantiifolia*) Lime trees are frost sensitive, small and leafy, and growing to 3 metres with narrow winged leaves to 8cm; these can be long blunt or rounded at the apex. Branches have sharp spines. Limes are a smallish (3–5cm wide), generally round fruit with yellow to green thin skin at maturity.

'Tahitian' lime or 'Persian' lime: (*Citrus × aurantiifolia* syn *Citrus latifolia*) is the most cold tolerant of the limes and, in Australia, is commonly grown as far south as Melbourne and can even be grown in Hobart in southern Tasmania, in warm sheltered places. Limes need a sheltered position, fully sunlit throughout the day and protected from cool winds. This tree grows to medium height but in cool regions will remain small. The thin-skinned fruit is small, rounded and best picked when green coloured. When they ripen and become yellow they often show stylar end rot so should not be left on the tree after maturity. The flesh is a greenish colour with few seeds. The juice is acidic with a distinctive 'lime' flavour and is sought after for drinks and for salads and cooking generally.

The 'West Indian' or 'Mexican' lime (*Citrus × aurantiifolia*) tree has sharp spines and grows to 3–4 metres if grafted to Trifoliate orange (*Poncirus trifoliata*) rootstock, but taller when grafted to other rootstocks. This tree likes tropical to semi-tropical conditions as it is intolerant to frosts. The leaves are aromatic and the fruit is small, rounded and pale green at first, turning light yellow at full maturity. Fruit, produced year round, has greenish flesh that is very acidic, aromatic and full of 'limy' flavour. Fruit is used mature when it may have green or yellow skin.

Rangpur lime (*Citrus limonia*) The Rangpur lime is thought to be a hybrid between an acid lime and a mandarin and is actually sometimes named a Mandarin lime. This small tree has dull green foliage and branches with some thorns. The yellow-orange coloured fruit is small oval or rounded in shape, sometimes with a short nipple. Flesh is orange coloured and acidic and can be used as a substitute for limes or lemons. The tree is cold- and salt-tolerant, crops year round and is an ideal pot plant. The 'Kusaie' lime is a small-fruited form of Rangpur lime and is also available.

'Rangpur' lime

Sweet lime (*Citrus Limetta*) This is a small tree growing to about 2m mainly suitable for tropical regions. It produces smallish fruit with a green skin and orange flesh (pulp). The fruit is used as an orange because the flesh is edible and non-acidic.

Limequat (hybrid of *C. aurantiifolia* × *Fortunella margarita*) Limequats were bred by Dr W. T. Swingle of the US Department of Agriculture in 1909. He developed three cultivars one of which is available in Australia: 'Tavares' is an original hybrid of (*C. aurantiifolia* × *Fortunella margarita*). The fruit is a light orange colour, small and necked, with acidic juice and edible skin. Limequats make an attractive foliage plant with some thorns, suitable for growing in pots.

Grapefruit and other large citrus fruit

Most of the citrus included in this group have large fruit. Some cultivars are mild tasting while others are very sour or acidic.

Grapefruit (*Citrus × aurantium* syn *Citrus × paradisi*) The grapefruit seems to have originated in the West Indies and is thought to be a hybrid or mutation of the 'Shaddock', introduced there earlier by a Captain Shaddock, a 17th century English ship commander. Its name probably comes from the fact that the fruit hang in large bunches. The modern grapefruit was introduced by the Spaniards to Florida, USA, in 1823.

Grapefruit has also been called 'Pomelo' (confusing it with Pummelo *Citrus maxima*), 'Small Shaddock' and 'Forbidden Fruit'.

CITRUS SPECIES AND CULTIVARS

The grapefruit tree is large (10 metres or more) with large winged leaves. It prefers hot sunny frost-free conditions to produce sweet tasting fruit. It will grow as far south as Tasmania if given a warm position, wind protection and well-drained soil.

Grapefruit is large (up to 15cm diameter and sometimes more), rounded and slightly flattened in shape, with light yellow skin, thick or thin depending of the cultivar. Some have pink- or red-coloured flesh; some are sweet or sour with or without many seeds. If picked too early or when grown in very cool conditions, grapefruit can be acidic and slightly sour. If left on the tree too long in warm areas there is a risk of dry stem, fruit drop and attack by fruit flies.

Grapefruit cultivars include 'Flame', 'Henderson', 'New Zealand Sweet', 'Ray Ruby', 'Red Blush' ('Ruby'), 'Rio Red', 'Star Ruby', 'Texas Sweet', 'Chironja' (see below), and 'Thompson'. Two main cultivars are in use in home gardens in Australia: 'Marsh's Seedless' from the USA, and 'Wheeny', an Australian selection.

'Marsh's Seedless' (*Citrus aurantium* 'Marsh's Seedless' syn *Citrus × paradisi* 'Marsh's Seedless'). This grapefruit was discovered as a seedling tree in Florida USA and was first listed as the 'Marsh' grapefruit and has since also been named 'Marsh's Seedless' because it is seedless as are the sports chosen from its progeny. Sports from the original tree have produced other cultivars such as 'Thompson', a pink-fleshed grapefruit, and, from a sport on 'Thompson' that arose by bud mutation, a red-fleshed cultivar 'Ruby'; both of these are grown in Australia.

'Wheeny' grapefruit, Mildura, Victoria

'Wheeny' (*Citrus × aurantium* 'Wheeny' syn *Citrus × paradisi* 'Wheeny') originated as a selection from seedling trees grown by R. J. Benton at Wheeny Creek, Kurrajong, NSW. The fruit is large and thin-skinned with many seeds and a sharp taste. It is a very juicy, heavy cropping type. Unfortunately, it does tend to be biennial bearing but in the 'on' year produces huge crops of fruit that hang in bunches. Pruning at the end of a light crop year will help to even out cropping.

The Poorman orange (see above) is sold in New Zealand as a grapefruit but is thought to be a hybrid between the 'Seville' and a sweet orange or a natural hybrid of the 'Shaddock' and not a true grapefruit. It is suitable for cool areas.

'Chironja' is a grapefruit-like tree (chance hybrid) producing medium to very large lemon-coloured fruit that is easy to peel. The juice is sweet and mild with no bitterness and this tree according to Morphett and Tolley[20] is a better choice for cool areas than the 'Marsh' grapefruit cultivar.

Pummelo (*Citrus maxima* syn *Citrus grandis*) This giant of all citrus fruits, is thought to be native to southern China, Malaysia and Thailand. It has been spread to all parts of the world but in Asia it has become a commercial crop and many cultivars have been developed.

The salt-tolerant pummelo tree can grow to 7 metres, and is best suited to tropical or semi-tropical areas where fruit will develop full sweet flavour. Trees will grow in most soil types and are relatively cold tolerant but, when grown in cool areas, the fruit may not be as sweet. I have seen trees growing successfully in the Mildura area of northern Victoria but they do not do well further south in Australia.

Pummelo branches have blunted spines or none at all. Leaves are large (up to 20cm long), oval and broadly winged. Pummelo fruit may grow nearly as large as a basketball and weigh several kilograms, although some cultivars have smaller fruit about the size of grapefruit. Fruit is globular or oblate and usually yellow-green at maturity. Flesh is usually coarse and grainy and can have a sour or sweet taste.

A pummelo has a very thick aromatic rind and white to pinkish pith under the skin and around each fruit segment. It is better to peel the pith away before eating. Although it is sometimes hard to remove the skin and pith, the flesh once uncovered is easy to separate into segments, and

Pummelo fruit, Guilin market, China

Pink form of Pummelo, Chinese market place.

usually very juicy and pleasant to eat. With some cultivars only about one quarter of the whole fruit by volume is edible flesh.

Both pink and yellowish-fleshed cultivars are grown. While in China, I tasted many different cultivars of this fruit, some with pink flesh, all with a very pleasant flavour without acidity.

Shaddock (*Citrus maxima* 'Shaddock') This is another name for Pummelo, referring to Captain Shaddock (see above) who was responsible for introducing seed of *Citrus maxima* into the West Indies from where it spread to other parts of the world.

Citrus grown for their leaves

Kaffir lime (*Citrus hystrix*) 'Kaffir' lime (also known in some other parts of the world as Leech lime or Makrut lime) is a species of citrus grown mostly for its leaves. The leaves are very aromatic and are used in preparing curries, soups, fish dishes and other delicacies and can be harvested all year round.

The leaves are distinctively 'waisted' and look like two separate leaves joined together. The fruit is very bubbly and lumpy and not generally eaten, but drinks have been made from it and the skin is sometimes

Aromatic Kaffir lime tree leaves showing the double-segmented leaf

candied. Juice and oils from the fruit are used in shampoos and cleaners with recognised germicidal qualities.

The small tree has large sharp thorns. 'Kaffir' lime trees prefer tropical conditions but can be grown as far south as Melbourne if protected from cool winds and given maximum exposure to sunlight.

Australian citrus species

Aboriginal Australians were well aware of the native citrus species, but until the late twentieth century, settler Australians showed little sustained interest. Species of citrus had been found growing in Australia mostly in rainforest areas or as shrubby species in areas that have been cleared by logging or farming activity. F. M. Bailey, the Colonial botanist of Queensland, described these species under the genus *Citrus* in his book *Comprehensive Catalogue of Queensland Plants* (1909). He described and gave rough drawings of *Citrus australasica* (Finger lime), *Citrus inodora* (Russell River lime), *Citrus australis* (Native orange), and *Citrus Garrawayi* (Mount White lime). Bailey is accredited with specifically naming *C. Garrawayi* and *C. inodora*.

These species were later reclassified from *Citrus* species to *Microcitrus* species. Elliot and Jones[21] describe the five species as *Microcitrus australasica* (Finger lime), *M. australis* (Wild lime, Round lime), *M. garrawayae* (Mt White lime), *M. inodora* (Russell River lime) and *Citrus glauca* syn *Erimocitrus glauca* (Desert lime). They have all now been reclassified once again into the genus *Citrus*.

Seeds of Australian citrus species were sent to America during the early 1900s and crosses made with other citrus species. In 1911, at an experimental citrus breeding station in the USA, a hybrid fruit named a Faustrimedin was produced from a Finger lime × Calamondin cross. Seeds of the Faustrimedin, Faustrine and Eremolemon fruits (other hybrids) were later imported into Australia, and breeding programs begun by Dr Stephen Sykes of the CSIRO Plant Industry have produced commercially accepted cultivars.

Now, selected cultivars and other hybrids are being experimented with. Cultivars suitable for home gardens and for the horticultural industry are being selected from wild populations. Dr Sykes has selected cultivars from the wild and from specific breeding programs and these include 'Australian Outback lime', 'Australian Sunrise lime' (Finger lime × Calamondin), 'Australian Blood lime' (Mandarin × Desert lime). Plants have been propagated for commercial plantings some of which are starting to bear crops to supply the growing demand for Australian bush foods. The next generation of crosses (F2 hybrids) with pummelo have produced two extraordinary plants: 'Son of Blood' and 'Aussie Pink lemon'. These and other cultivars, such as the Red Finger lime selection 'Rainforest Pearl' (PBR), will soon be available to commercial growers as will selections from the Australian Finger Lime Company: 'Purple Viola', 'Jali Red', 'Mia Rose' and 'Alstonville Pink Ice'(all PBR registered). Some of the Australian citrus species are now being used as rootstocks for other citrus cultivars.

Trees are usually spiny, small growing and have tiny fruit, many of them with a bitter or sour taste and hardly any juicy flesh. All fruit of these species is edible, some are rather dry but others very juicy and there is a range of seedling variation. Some have blood red flesh. Some of the trees found in wild populations produce sweeter fruit with more juice and flesh (pulp). Selections with a larger flesh to skin and pith ratio are ideal for use as fresh fruit or for juice.

Australian citrus species are ideal for nearly all well-drained soils. They tolerate stress (including some drought) and will grow just about everywhere in Australia if protected from winds and frost in cooler areas. They are excellent for growing in small spaces including as potted plants, plus they supply bite-sized unusual delicious fruit.

The Desert lime or Native cumquat (*Citrus glauca* syn *Eremocitrus glauca*) is of particular interest. It is found in arid to semi-arid regions in Qld, NSW and SA, often in suckered plant clusters. Branches are prickly–thorny and leaves are small (1–5cm long), thin and boat-shaped often with a notch at the leaf end. Sometimes during harsh weather conditions in the arid environment, plants become leafless. Plants have tiny white flowers and small thin-skinned fruit (up to 2cm diameter), lemon yellow when ripe. Fruit is juicy with an acidic taste. Selections of plants with milder fruit are currently being undertaken, as are trials for use as a rootstock for other citrus species.

- New citrus cultivars are being registered in Australia all the time. Some are parthenocarpic (able to produce fruit without seed). New cultivars include the following:
- 'Allen Eureka' lemon, the most popular lemon grown in California, virtually thornless and a very high quality fruit.
- 'Eureka Seedless' lemon, originating in South Africa. It is totally seedless with all the characteristics of the original 'Eureka' selection.
- 'Eyles Kaffir' lime: a selection of the 'Kaffir' with very large aromatic leaves and reduced thorniness.
- 'Seedless Eureka' (syn 'Eureka' lemon): a lemon cultivar from 2PH Farms, Emerald, Queensland (variety access) and is from an irradiated selection. The fruit is seedless with a very smooth skin.

(Source: *Good Fruit and Vegetables*, Rural Press, December 2003.)

CHAPTER 3

propagation of citrus

Citrus trees can be propagated in various ways: by using selected seed, aerial layering, cuttings, root sections or by budding or grafting onto selected rootstocks or existing trees. Generally, the success of propagation of citrus using cuttings is rather poor by commercial standards as the plants have no acceptable pathogen resistance. Some citrus do, however, propagate easily from cutting and can be used to supply potted plants for the home gardener.

Commercial propagators using micro propagation techniques of tissue culture in a laboratory can now produce thousands of plants from one tiny piece of plant material. In-vitro grafting onto tiny plants with stems of only 2–3mm thickness is also used to eliminate viruses or to provide very tiny grafted plants.

This chapter will detail methods of propagating citrus, paying particular attention to those most easily available to home gardeners. I suggest that home gardeners try their hand at chip-budding, micro-budding, bark grafting, aerial layering or propagation by cuttings.

Damian Ridgewell with 'Lane Navel' oranges and huge regrowth leaves taken from a recently grafted (reworked) tree

Propagating citrus by seed

Citrus trees were initially propagated and spread by seed. Seeds were easily carried (usually within fruit so that the seed did not dry out) and could withstand the extremes of long journeys.

Propagating Trifoliate orange seed for rootstock growing organically, using compost as a medium

Young seedling citrus trees in tall pots necessary for developing non-tangled root systems

Some seedling trees grown to maturity bore different fruits to those of the parent plant and these added to the genetic mixing of the known citrus cultivars.

Citrus can be propagated easily from seed. The fresh seed should be sown about twice the depth of the thickness of the seed and kept moist in deep pots that must be given 40% shading if grown in the open. The seed has no need for cool storage before sowing because they have no dormancy period. I have seen seed of a citrus fruit left on the tree too long germinate inside the fruit. For short-term storage, the seed can be stored in a container with material such as moist copra peat.

Commercial propagators treat seeds by heating them to 53°C for two hours; they are then chilled until needed. Many of the seeds of citrus species such as limes and lemons are polyembryonic and have the ability to produce more than one plant from one seed, most of them genetically identical to the parent plant and these develop from nucellar embryos. Some of the plants that develop from the seed, however, may be the result of pollination and have mixed genes from two parent plants. Citrus such as citrons and pummelo produce seeds that have embryos that are the result of pollination only and do not produce nucellar seedlings; as each of their seeds contains a single embryo, they are called mono-embryonic.

It is not advisable for home gardeners to propagate known cultivars of citrus trees from seed and grow them to maturity because of the variation that may occur. Seedling trees often have very spiny, armed branches and they can take many years (up to 30) to become fully fruitful. To obtain a specific cultivar, it is best to propagate by budding or grafting (see below). However, it is useful to know how to raise seedlings for use as rootstock plants as rootstocks are not always readily available. Seed from fruit of the sour orange, Rough lemon, mandarin and Trifoliate orange, for example, have been used to grow rootstock material.

Home gardeners wanting to propagate from seed should use seed collected from fully mature fruit. This should be washed and dried, then refrigerated until needed (for a few weeks) or it can be planted out immediately. Only plump seeds should be chosen and they must not be allowed to dry out completely as this is detrimental to seed germination. Seeds can be planted singly in pots, or in threes (the two smallest seedlings being removed at a later date). Seeds can be planted *en masse* in pots or trays for later transplanting into pots or soil. Use an Australian Standards compliant mix.

For planting out seedlings, pots should be selected that will not encourage roots to tangle or twist. Seedlings should be planted into a registered potting mix. If seedlings are planted in rows, they should be about 10cm apart.

Once planted, seedlings are susceptible to drying out, so they require regular watering and feeding to encourage them to grow quickly to a size suitable for budding or grafting; this can take from one to three years. Small seedling trees can be grown in pots under shelter for micro-budding (see below) during the summer or autumn.

Sucker growths, spines and side branches should be removed from seedling trees as they grow to produce a single straight trunk on which to bud or graft. A tree with many side branches or spines will have no clear trunk space for bud insertion or grafting.

Aerial layering

Citrus trees can be propagated by aerial layering or marcotting. Still attached to the tree, branches selected for layering are usually 5 to 10mm in diameter but larger limbs can be used.

Begin by cincturing the selected branch, that is, by removing some bark from all around one point on the branch (see also Glossary). Use a clean pair of pliers or multigrips, budding knife or other cutting equipment. The bark is completely removed in a section 1–2cm long at a time when the bark tends to slip easily, usually early spring or late summer. Root promoting hormone powder, liquid or gel or an organic substance such as pure honey can be placed onto the injured area for quicker root promotion.

Once the bark is removed and the wound area prepared, a handful of moist but not wet sphagnum moss or other material such as coconut fiber (copra peat or palm peat) is wrapped around the injury point, and aluminium foil or thick plastic wrap placed around that. This wrapping is then tied into position to seal it.

The aerial layer should be checked regularly for moisture content and to see if any roots have developed from the wound area. If clear plastic wrapping is used to cover the aerial layer there is no need to unwrap it to examine roots. Once roots have grown and filled the wrapping, the branch can be severed just below the bottom of the aerial layer, the plastic removed and the branch with roots attached placed into a pot.

I have seen an injured trunk of a lemon tree produce roots where it had clay wrapped around it. However, using clay as a root promoting material for aerial layering is not usually recommended as it may contain soil organisms, harmful bacteria or fungi that can cause rotting to occur around the injury point.

Propagation from cuttings

Citrus trees can be grown from cuttings using one leaf and an attached short stem section (leaf bud cuttings), small pieces of mature stems (semi-hardwood cuttings), soft new shoots (softwood cuttings) or root cuttings (root sections). Most citrus, however, are usually budded or grafted onto a seedling or a selected rootstock.

Material for cuttings can be selected by choosing shoots from the donor tree that receive full sunlight and in an area where fruiting is established. The cuttings can be stored for a short time in sealed plastic bags that have been misted inside with water and the bags placed into an 'Esky' or in the crisper section of a refrigerator until ready to use. The same storage principle applies for shoots (scions or budwood) chosen for grafting purposes (see below). Some citrus species form roots readily when grown from cuttings but others may take a longer time (6–12 months or possibly more).

Leaf bud cuttings are made by cutting selected shoots into sections containing one leaf and a short section of stem (5–10mm) above and below the leaf. Sometimes the lower piece is cut longer than the top piece to enable it to be stuck deeper into propagation mix. These cuttings are stuck into propagating mixes using perlite, sand, synthetic propagation blocks or copra peat in trays. It may be advantageous to apply root-promoting hormone or honey to the base of the cuttings.

Leaf bud cuttings must not be allowed to dry out so they should be placed in a sealed propagating box or something similar to ensure success.

Semi-hardwood cuttings are cuttings taken from the base of shoots when they have just stopped growing after a growth flush and have no new leaves forming on them, or cuttings from the matured (lower) sections of laterals that do not stop growing.

Cuttings are usually between 10 and 30cm long and leaves are usually stripped from the lower part of the stem for pushing into propagating medium. Another approach is to remove all leaves then completely wrap the bare cuttings in film wrap (e.g. Parafilm™) leaving a part of the lower end that is to be to be pushed into the propagating media wrap-free.

Cuttings are sometimes wounded on their base by scraping off bark on one side for 1–2cm and applying a root promoting substance. The cuttings are placed into propagating mix in trays or pots and watered. The containers are put into a cold frame or hothouse. The plastic film wrap is removed only when the cuttings have rooted.

For those without access to a hothouse or cold frame, hothouses for small cuttings can be improvised using old boxes or by building a small frame and covering with two layers of plastic sheeting or bubble plastic and then sealing the gaps. A large pot with a bent wire frame covered with plastic will also do provided that the plastic does not touch the cuttings. The plastic sheeting must be wrapped completely around and under the pot for a complete seal. Heat and moisture will be trapped inside creating high humidity for cuttings. Less watering will be required after the initial watering. I have seen sealed pots left without further watering for as much as five weeks; by that time cuttings had already formed roots.

Softwood cuttings are small cuttings 5–10cm long, taken from the top of actively growing shoots. They are soft with new leaves sprouting from their tops. They are difficult to handle because they can wilt very quickly. If the cutting is too soft and to prevent wilting, the active growth section at the top can be removed. This may not be necessary if cuttings can be placed very quickly into a high humidity environment such as a sealed propagating box. Where available, a constant misting apparatus can be used. If cuttings are prevented from wilting they will root readily, especially if the base of the cuttings is kept warm.

Root cuttings were used for propagating citrus during early settlement before rootstocks were readily available. They were also used for propagating from a selected seedling that showed promise as a commercial variety. Cuttings are taken by cutting pieces of roots into sections 10 to 30cm long. The larger diameter end is pushed upright into propagating mix with the top 1–2cm exposed. The section exposed to sunlight produces shoots and the base forms roots. This method is not, however, recommended for home gardeners because of the likelihood of transfer of diseases and because some trees can be susceptible to root rot.

Budding and grafting

The most common and reliable way to reproduce identical plants is by budding and grafting. Successful cultivation of most fruit tree cultivars relies upon the fact that they can be propagated by the transfer of a bud

Budding into new growth on cut stump of citrus tree; note slats used to support new growth.

Multi-grafted citrus tree

or buds, or a piece of growth containing several buds (scion), from one tree to another (rootstock), producing identical trees growing the same fruit. Commercial orchards depend on various methods of budding and grafting for the establishment of orchards able to produce fruit with consistent characteristics.

Home gardeners can develop good budding and grafting skills by following a few simple guidelines and by practising the techniques detailed below.

Equipment

Knives and secateurs are most important and must be very, very sharp and clean. Clean-cut edges and quick callusing (healing growth that covers wounds) at the graft point help prevent disease contamination. To check for sharpness, lightly run the cutting edge of a knife or blade along a sheet of newspaper, or use the usual cutting action with secateurs. If the result is a clean cut, the blade edge is sharp enough for use; if the paper is rough cut or torn, resharpen. Knives and secateurs should be cleaned and sharpened regularly.

To cut buds a very sharp budding knife is recommended, but one-sided razor blades, or extremely sharp kitchen or pocket knives, can be used. It is important, though, to sharpen only one side of the blade on any cutting tool as a V-shaped cutting edge can rip bark and make ragged edges. This may cause poor grafts to form.

Once the graft is done by whatever method, completely seal it at the graft point by wrapping with budding tape or cling wrap, or by sealing with grafting wax. Tape or cling wrap are removed once the graft has taken. These tapes or ties are important for holding the cambium layer of the scion and rootstock together so that they knit properly. The tape or graft cover also prevents moisture penetrating the graft and helps keep out harmful bacteria and fungi that may cause disease or rot.

I have been developing the use of a narrow plastic sleeve over the graft scion, or chip bud and graft union area to help ensure grafting success. The top of the sleeve is sealed and the bottom is open-ended to allow air to circulate within it. Before placing this sleeve over a graft union area, use a spray bottle to mist the inside with water. The plastic sleeve acts as a humidifier and mini-greenhouse, preventing the graft from drying out and protecting it from extreme weather conditions. Plastic sleeves also protect the graft from insects, animals and weather and keep the graft warmer, which in turn encourages quicker callus formation at the graft union.

The plastic sleeve can be made from tubular packaging material manufactured for packing small items, inexpensive and available in large rolls. It is even more economical to cut and stitch up recycled clear, white or opaque plastic to make into narrow plastic bags for the purpose. Plastic sleeves can also be made from bubble plastic for extra warmth. Stretch Parafilm™ (a medical self-sticking wrapping plastic that can be moulded around the scion) will also work well in conjunction with the plastic sleeve.

Anchor the plastic sleeve below the graft union area with a pin or piece of grafting wax to prevent it blowing around. It can be left on until the graft piece starts growing, or until new shoots reach the top of the inside of the sleeve. In very hot weather, the enclosed leaves on the graft piece may burn (especially when scions with leaves still attached are used) so shade the graft by covering the plastic sleeve for a short time with a paper bag or partially painting the outside of the sleeve with white paint.

Rootstocks

Rootstocks are very important to the successful growth of budded or grafted trees. Some rootstocks used for citrus have a dwarfing effect on some cultivars. Others gradually develop viral symptoms, for example, Scaly Butt can develop after a 12-year period on trees grafted to Carrizo citrange, Rangpur lime, Troyer citrange and Trifoliate orange rootstocks. Several citrus species are incompatible with particular rootstocks so will not graft properly or the tree will die soon after grafting. It is important for commercial citrus or nursery growers to use virus-tested budwood and rootstocks for grafting purposes (see Auscitrus in Glossary and Chapter 1). Most rootstocks used in Australia are tolerant to both Tristeza virus and Exocortis virus but all are susceptible to Psorosis virus.

The rootstock used can affect tree size, ripening time, cold weather tolerance, and resistance to pests and diseases, particularly root diseases and viruses (see Chapter 6). Quality and quantity of fruit can also be affected as can sweetness or sourness, juiciness, colour, and thickness of skin. Some rootstocks prefer light sandy soils; others are suitable for heavier soil types and resist waterlogging. Different countries have adopted different 'favourite' rootstocks for their regions. Following is a list of some citrus rootstocks, their uses, and some of the advantages and disadvantages.

Different rootstocks	Description and uses
Flying dragon (*Citrus trifoliata* var. *monstrosa*)	Very thorny and hard to work with but does have a dwarfing effect by delaying spring growth and inducing early dormancy, producing small trees suitable for small garden areas and for pot grown plants producing excellent quality fruit. Tolerant to Tristeza virus and nematode-resistant, but susceptible to Exocortis and Psorosis virus. Suitable for all citrus except Imperial mandarin.
Citranges including Carrizo and Troyer citrange (*Poncirus trifoliata* × *sinensis*) crosses	Used for most citrus cultivars and perform well. Resistant to root rot and particularly useful when replanting old citrus groves where soil has not been fumigated to kill nematode populations. Fruit produced using this rootstock tends to have higher quality and more juice compared to other rootstocks. Budwood used for grafting and budding, these rootstocks should be free of Exocortis virus. Eureka lemon is incompatible with these two rootstocks so unsuitable for budding and grafting. However, Benton citrange (*Poncirus trifoliata* × *Citrus sinensis*) bred in Australia by crossing Trifoliate orange with 'Ruby' blood orange, is compatible with Exocortis free Eureka lemon scions and budwood, and is used where these two citranges cannot.  *Poncirus trifoliata* fruit and seed
Citronelle or Rough lemon (*Citrus jambhiri*)	Used for citrus grown in free draining soils. Slightly drought tolerant but imparts less juiciness, early ripening and less sugar content to fruit. Does not perform well for many mandarin cultivars. Citronelle is a valuable rootstock for lemons that are grown in dry areas or on sandy soils, although it is subject to collar rot and root rot in wet soils. Trees bearing rough lemons are often seen in home gardens where the rootstock has outgrown the grafted tree or when rootstock sucker growth has not been removed.

Different rootstocks	Description and uses
Mandarin 'Cleopatra' (*Citrus reticulata*)	Especially used for mandarins. Mandarin cultivars have good results as a rootstock for citrus trees in 'heavier' soils (i.e., containing more loam or clay and less sand). Fruit are smallish but of good quality. Moderate resistance to root rot.
Rangpur lime (*Citrus limonia*)	A fair degree of salt and lime tolerance and Tristeza virus resistant. Buds or scions used to graft onto these rootstocks must be free of Exocortis virus (Scaly Butt). Fruit are good quality and trees bear heavy early crops compared to other rootstocks.
Sweet orange (*Citrus sinensis*)	Grafting to sweet orange rootstock usually results in large, productive trees with high-quality fruit, especially in areas where no other citrus have been grown previously.
	Used for all common citrus species. Seems to be free of virus infection symptoms (tolerant), although it is susceptible to root rot (*Phytophthora citrophthora*). Make sure trees do not become wet, waterlogged or watered with saline water and keep mulch or grass away from direct contact with root or trunk area to minimise risk of root rot.
Sour orange syn Seville orange or Bitter orange: (*Citrus aurantium*)	Once used for all citrus species but very susceptible to Tristeza virus so it is now restricted to lemon and lime trees only. Tolerant to root fungi, cold and drought resistant and produces high quality fruit.
Trifoliate orange (*Poncirus trifoliata*)	A deciduous rootstock especially useful in cooler areas. Some tolerance of moist conditions and some resistance to root rot (*Phytophthora citrophthora*). It will grow in most soils, but will not tolerate highly alkaline or saline soils. Must be watered regularly if grown in free draining sandy soils. This rootstock is ideal for pot grown plants.
	Fruit is of very high quality, very juicy with highly coloured skins. If Exocortis virus is present in the budwood or scions, resultant trees will be dwarfed and not very productive so use virus-tested material.

Note for home gardeners: To avoid serious viruses in propagating material purchase citrus trees from registered reputable nurseries. Use these trees for any propagation material needed as most viruses are transmitted by using propagation material from infected plants.

Scions

'Scion' is the name given to a piece of the bud stick chosen for grafting and, as with rootstocks, the choice of scions is important as is their care.

The plant material used for budding and grafting can influence success or failure. Scions or shoots for buds, should be cut from the outside edge of the donor tree where shoots are fully developed, the stems rounded in shape and exposed to sunlight. Young vigorous shoots often have a triangular cross-section and are not as suitable for bud sticks (but can be used for micro-budding) because of the lumpy shape and less cambium contact area on buds cut from these pieces. Not so long ago, all citrus budding was done on well-developed seedling trees, using budwood of mature shoots well rounded in shape. Buds from the scion were large, shield shaped, and a large seedling tree with a thick trunk was needed to accept the bud. Now with micro-budding (see below), much smaller buds can be used so that rootstocks stems can be younger and much thinner.[22]

With plants that are difficult to graft (e.g. with thin-stemmed growth) it can help to ring-bark below the shoot chosen for bud sticks for a short period before collection; this builds up plant sugar reserves in the shoot and can give better grafting outcomes. Also helpful is to reduce the amount of light to the plant (this is called 'etiolation') for about 3–4 weeks by using heavy (50% or more) shade-cloth covering. The plant is then allowed full sunlight for a week before taking cuttings or graft material.

On a scion 10–20cm long, there are usually 5–10 buds. These scion pieces have leaves attached when taken from evergreen tree species. Sometimes, any leaves attached to the scion to be grafted are cut in half or removed. This is not essential, however, if a plastic sleeve is used to cover the graft (see above).

Budsticks are usually cut from the donor plant when needed but they can also be stored for a week or more inside a sealed, finely water-misted plastic bag in the crisper section of the refrigerator (or in an 'esky' or other cool dark place). Budsticks for budding and grafting can be cut into 2–5 bud pieces (scions) if necessary.

Buds for budding are found at the base of leaf stalks and may be well formed and easy to see or very tiny and hardly visible to the naked eye. Buds are selected from healthy budwood and are cut from the stick with a scooping action using a budding knife or sharp knife. Starting 1–2cm above the bud to be cut and using a long cutting action, cut 2–3mm into the bark downwards and behind the bud to 1–2cm below the bud then up again. Some propagators actually cut the buds from the bud stick and retain the leaf or at least part of the leaf stalk to act as a handle when inserting the bud under the bark of the rootstock. For microbudding into very tiny rootstocks, tiny bud dimensions are very much smaller and the bud cutting action different (see below).

If virus-resistant rootstocks compatible with most citrus species are used (see the table 'Different rootstocks – Description and uses', above), then three or four different buds of different species can be placed in a spiral configuration around a medium-sized rootstock trunk. Some citrus such as the citron will grow at a much faster rate than say a cumquat, so if grafting one citrus to another, particularly a strong grower to a medium growth rate tree, the tree may overbalance unless the strong branch is supported to prevent it breaking from the tree. Where there are unequal growth rates, the graft area can produce a mass of callus overgrowth that can break off.

Different grafting methods are used for different species, and often the method used depends on the time of year and type of graft material (budwood) available. Grafting is traditionally done during spring or late summer, and budding in spring, summer or late summer.

Budding

Budding is actually a form of grafting. It involves lifting the bark of a small branch or lateral and placing a live bud under it so that it forms a graft, or inserting a chip with a bud into a slot on the rootstock. Once the bud has taken, the limb is cut back to the inserted bud, causing the bud to grow and to eventually produce identical fruit to that of the parent plant.

'T'-budding

Along with chip-budding, this is one of the most used methods of citrus propagation and involves the insertion of a small piece of bark (with a bud attached) under the bark of a rootstock plant. Budding is usually done in early summer or autumn but can be done anytime the bark is able to 'slip' easily to allow the insertion of the bud. If the bark does not lift easily then budding is not recommended. The rootstock must be healthy, and it is a good idea to feed and water it well a couple of weeks before budding to make sure the sap is flowing.

New growth on bud recently inserted into cumquat

Although the normal 'T'-bud insertion method will work, the inverted 'T'-bud cut is prefer ed for citrus as this seems to give better results and is less subject to water entering the wound area. Two cuts are made on

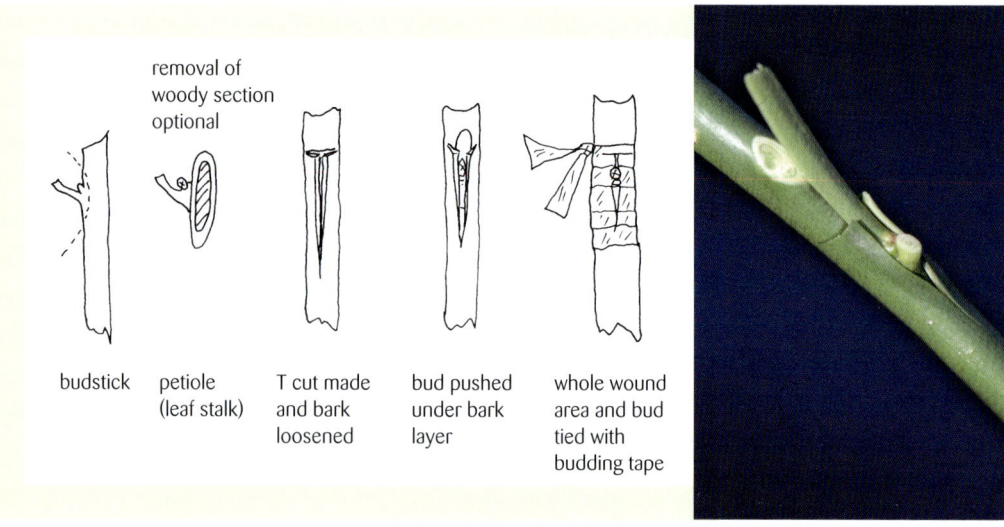

'T'-budding in citrus lateral: the long 'tail' is removed after bud insertion

the trunk of the tree to be budded. One cut is horizontal, the other vertical and they join to make an inverted 'T'-shaped cut in the bark. The vertical part of the 'T' must be long enough to insert the cut bud so this will vary according to the diameter of the bud stick from which the buds were cut. The 'T' cuts must be at least 30cm above where the trunk enters the soil, to prevent fungi or bacteria being splashed onto the budded region.

Making a slight chisel cut at the end of the bud piece to be inserted will greatly enhance the ease of placing the bud under the bark. The 'hump' of a budding knife or the knife edge is used to lift the bark along both sides of the vertical cut of the 'T' made in the rootstock so that the bud can be inserted by pushing it downwards, or upwards for the inverted 'T', to slide along the slit until it is well seated and the bud is situated at about the middle of the vertical cut. Make sure to place the inserted bud the right way up; if it is inserted upside down it will not grow.

The 'T' cut area with its inserted bud is then tightly covered with budding tape; this helps press the bark together aiding cambium contact between bud and rootstock allowing the bud to 'take' (graft) properly. Some propagators leave the actual bud exposed and just tape all around it. Budding tape is usually left on for 3–5 weeks or until it is evident that the bud has taken or has started to grow. Once the inserted bud has grafted successfully, the tape can be removed, and the rootstock can be pruned off above it. Alternatively, the rootstock can be cut halfway through just above the bud and the branch bent to one side while still attached; this will help the inserted bud start growing, and is the best practice when large seedling trees are budded. Once the inserted bud has started to grow the bent rootstock piece can be pruned off. If the limb or trunk is not bent to below the inserted bud or pruned off, the bud will not be forced into growth. I have seen an inserted grapefruit bud on a lemon tree branch remain dormant and inactive for many years and not actually start to grow until the limb was cut off near the bud; once this was done the bud developed well.

Some grafters when grafting seedling trees or small branches leave a 10–20cm piece of rootstock above the budding site and remove any buds or shoots growing on it. Growth from inserted buds or from scion grafts can be very vigorous and subject to wind damage so the bare stub is used to tie or anchor the new shoot in place as it grows from the inserted bud. New graft shoots on larger trees can also be supported by loosely tying them to thin slats attached to the tree.

Chip-budding citrus

Chip-budding is so named because the process removes a small piece of wood and bark containing a growth bud from one plant and places it into an identical 'chip' recess cut into the rootstock plant. The chip piece is usually cut from previous season's lateral growth or from new growth that is still growing during the spring–summer season.

I have had success chip-budding with some plants all year round. Chip-budding does not depend on the bark 'slipping' before it can be done. Scions can be used to supply chip buds if needed and can be stored for two or three weeks before use.

A sharp knife or budding knife is used to cut the bud out of the lateral. Beginning about 1cm above the bud on the lateral (or just above a leaf on new growth) make a cut angled at about 30° downward behind the bud to just below the bud (1cm or less). The knife is taken out and then placed below the bud and a cut angled at 45° is made downwards through the lateral until it intersects the first cut. This creates a chip containing a bud, with a 30° and a 45° cut edge. This chip is transferred to a site (socket) identically cut on the rootstock (anywhere on the rootstock stem or branch). The chip removed from the rootstock is discarded.

If the rootstock stem and the lateral from which the chip bud was taken are the same diameter, when the chip bud is in place it will fit exactly and be hard to see. Once in place, the chip bud is taped (see Equipment above) and the tape left on for 3–5 weeks before removal. Some gardeners only loosely tie the chip bud into place at the top and bottom of the growth bud; some tie all the cut area but leave the growth bud uncovered; others cover the entire chip bud. I prefer covering the entire chip bud when budding is done in the field but this is optional.

Summer chip-budding into new growth, grafts quickly and the tape can be removed in 2–3 weeks if the chip bud has grafted properly. An inserted chip bud can be seen to have grafted when there is a white or cream callus growth around the cut edges of the inserted chip bud. Once grafted into place, it is safe to cut the rootstock back to the inserted bud, which will then grow into a tree or branch producing the same flowers and fruit as the parent plant. If the branch or lateral is not cut back to the inserted bud, the bud will not commence growth.

Chip buds can also be placed into large diameter rootstocks, but it is important that the edges of the cambium layer of the chip bud be aligned with the cambium layer of the rootstock, otherwise no grafting can take place.

To decrease the time taken for a chip bud to graft, I have used budding tape ties and on top of this I have wrapped strips of bubble plastic to increase the temperature and retain warmth: callus formation thus occurs more quickly. For extra warmth to the graft area, the rootstock can be cut back to 10–20cm above the inserted bud or 'T'-bud and a moisturised plastic sleeve (see above) can be placed over the rootstock lateral and down over the inserted chip bud or T-bud; this can be left on until the graft has taken. The sleeve can be

Chip-budding of 'Alfourer' mandarin into Carrizo/ Citrange rootstock

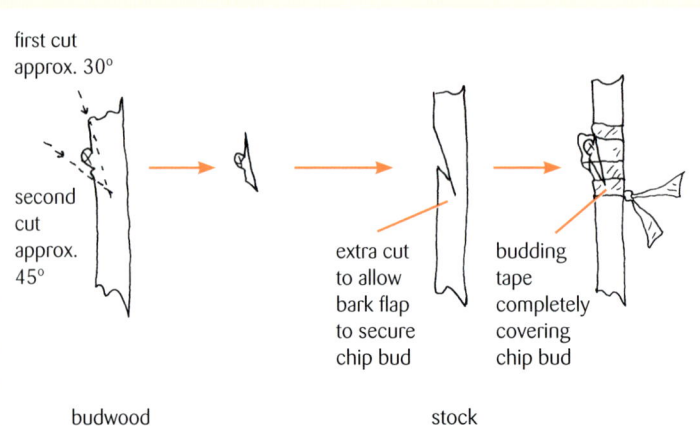

Chip budding

covered with a paper bag or partially painted with white paint if budding is done during hot weather or where UV levels are high (for example Tasmania). Alternatively, plants can be grown in pots under shade to prevent inserted buds from being sunburnt or stressed by excess heat. A plastic sleeve should be used when chip buds are chosen from new growth and the leaf of the chip bud has not been removed. I have found that the plastic sleeve helps the grafting process when using chip buds and chip pieces (large chip buds with one cut side placed under bark on a cut stem) with the leaf or part of the leaf still attached, even though the leaf inside the sleeve may eventually die.

If more than one cultivar is to be grafted to one rootstock, then several chip buds or T-buds can be placed into the rootstock but they must be placed in a spiral around the rootstock circumference so that no one bud is directly above or below another bud.

Summer chip-budding can be done into mature new growth or into laterals that have resulted from harsh spring pruning deliberately done to create lateral growth for budding into. To change fruit cultivars or to create a multi-grafted tree for variety and interest, the best way is to cut the tree branches to stumps in spring, allow new lateral growths to grow from these cut branches, thinning them out to about 10cm apart, and then to bud into the lateral growths during summer or autumn or even spring of the following year.

Micro-budding

This is almost the same as budding, except that smaller growths (bud sticks) are used for obtaining buds, and buds are placed into smaller diameter rootstocks or seedling trees or small laterals with a trunk diameter of at least 5mm where the bud is to be inserted.

Micro buds can be taken from mature shoots with well-formed buds. Bud sticks are, thus, young but mature shoots, either rounded or tri-angular (as they often are on young citrus shoots) in stem cross-section. Buds must be removed, using a very sharp knife preferably a budding knife, so that they have a flat base and are not 'scalloped' from the bud stick as with conventional budding removal action. It is unnecessary to remove 'wood' from the back of the bud as the bud is too tiny.

Buds can be stored for several days if necessary in sealed jars or plastic bags that have been moistened inside with water. Some gardeners place the cut buds in their mouth to keep them moist and to allow both hands to be free during the budding operation, but if you do this make sure you have good mouth and teeth hygiene, as I have heard of a propagator in the habit of eating very hot chilli for lunch every day wondering why his buds all failed to take. I suspect the chilli inhibited the buds in some way and stopped them grafting properly. There are many bacteria in mouths that may infect buds.

Buds are placed into an inverted 'T' cut. The bar of the 'T' is cut after the upright of the T, using a 'bowing' action: this lifts the edges of the bark slightly, allowing easy placement of the tiny bud into the cut by sliding the bud upwards, not downwards as usual with 'T'-budding. When the bud has been inserted, the area is fully wrapped with the bud covered. Ties are usually left on for 3–5 weeks before removal. Once the tape is removed, the rootstock or seedling is cut off just above the inserted bud and the bud will grow into a tree.

Budding can be done anytime the bark of the rootstock 'slips' easily, and this is usually between October (mid-spring) to early January (mid-summer) and late February (late summer) to mid-April (mid-autumn).

Micro-budding has a number of advantages. It is a faster operation that produces a tree in the shortest time possible. The tree trunk does not have a large wound area open to fungal or bacterial infection. Smaller trees are produced that are easier to handle. For those wanting a challenge, some people can bud and tie as many as 600 trees in a day!

Scion grafts

Scions with several buds attached can be used to change one cultivar of citrus to another or to add one or more cultivar branches to a tree. Scion grafts can be done on just one limb of a tree but usually, if a tree is to be 'grafted over', the whole tree is cut down to a stump leaving lengthy branch stubs into which bark grafts are inserted. All the bark on the stump should be painted with white, water-based paint to protect it from sunburn, especially as grafting using scions is best done during the warm summer months. Sometimes, one limb is left uncut as a 'nurse limb' and to provide sap flow and shading for the grafts. Once grafts have taken and are growing well, this 'nurse' limb can be removed and any sucker growth rubbed off the stump below the inserted grafts. If grafting fails, however, selected sucker growth can be allowed to grow for budding or grafting onto at a later date. If grafting with scions is unsuccessful, buds can be inserted into the strong lateral growths that emerge from the cut stumps giving home gardeners a second chance. Personally I prefer to cut the stump, allow selected branches to grow and bud or graft to these.

Scion grafting method of bark grafting of citrus

Bark grafts

Limbs to be grafted should be cut with a very neat, smooth cut at the end without any ragged or torn bark; re-cutting the limb to neaten may be necessary. Using a budding knife, the end of the cut limb should be slightly chamfered to make grafting easier. Citrus trees have very thin bark and it is easy to rip or tear the outer bark if you are not careful.

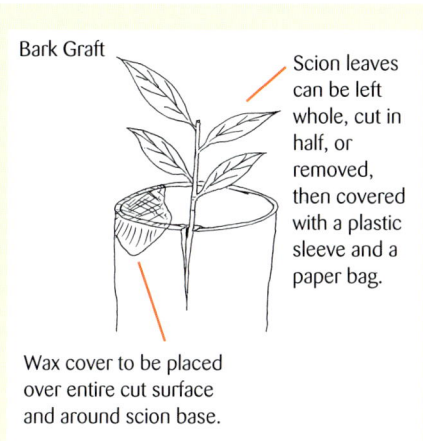

Bark Graft

Scion leaves can be left whole, cut in half, or removed, then covered with a plastic sleeve and a paper bag.

Wax cover to be placed over entire cut surface and around scion base.

The base end of the scion (see Glossary) to be grafted should be sliced at a 30° angle to create a sloped cut; the base of this is cut to form a tiny chisel end able to be gently pushed under the bark of the rootstock to a point where no part of the cut scion surface can be seen. The scion is attached with the cut of the sloped edge facing inwards towards the centre of the rootstock limb. Several scions can be placed evenly spaced around large limb stubs.

To prepare branches for grafting, using a sharp knife make downward slits 2–3cm long in the thin bark. A small sterile screwdriver or similar implement may be used to loosen the top of the slit for a few millimetres to help with scion insertion. Scions are then pushed gently under the bark along this cut. Sometimes scions are secured with tacks. The graft area can be sealed with grafting wax or budding tape, making sure to cover the cut limb surface and any splits in the bark as well.

In October 2005, I met Frank Lee, who seems to have tried grafting anything and everything constantly working out new grafting techniques. His mentor and friend, the late Wally Parkinson, taught him many techniques including the use of eucalyptus oil for cleansing grafting implements to ensure grafting success. Frank explained to me his special method of grafting onto citrus tree stumps or large branch stubs and I outline the method for readers. This method described has been used to change sweet orange trees to produce blood oranges but can be used for other citrus trees and other cultivars.

Frank's method of grafting citrus, depends on timing and must be done when the weather is warm with sap flow in the tree active. He relies upon quick action, sharp tools, cleanliness and patience. A tree or branch is cut to a stump and is cut so as not to leave any ripped or broken bark. This may mean that two cuts are made to ensure a neatly cut surface. The edge of the stump is immediately chamfered to bark depth with a sharp knife and scalding hot candle wax (grafting wax and bees wax could also be used) is spread thickly by brush to cover the whole wound area and a section of the stump below the cut. It is important that the wax be placed on the wound area and stump end as quickly as possible after the cut has been made. The stump and treated area is then covered with black plastic sheeting that is tied tightly to the base of the stump. Patience is now needed because the black plastic cover is left on for about 30 days before it is removed and bark grafting begins. (My view is that this 30-day time delay is consistent with callus growth beginning to form under the bark around the edges of the wound area which could be advantageous to successful grafting.) Bark grafts are then inserted around the circumference of the cut limb. The hardened wax is pushed aside, the scions are inserted and the soft wax moulded back around the graft area. After the scions are inserted an open-ended plastic bag or sleeve is inserted over the scion. A protective domed wire cage covered in green shade-cloth is then placed around the stump and left there until the scions start to grow which should be within 4–5 weeks. The shade frame and plastic cover are removed when the scions have started growth.

Frank Lee demonstrating grafting techniques

Frank says he has had success with inter species grafts using this method. The inter species grafts include: grafting a willow to a pine tree, a mandarin to a White Gum (*Eucalyptus* spp.) and a pepper (capsicum) grafted to a White Gum. I, myself, have not confirmed this aspect of his method but I am currently experimenting.

To keep grafts moist, some grafters completely wrap the scion in thin film wrap plastic (e.g. Parafilm™) before grafting, then cover the graft with a plastic sleeve and then a paper bag for shading and protection from the sun. Bags and film wrap are removed when the graft piece shows signs of growth. My plastic sleeve method (as already described) with a paper bag over it is equally effective.

Cleft or wedge graft

This is a very old method of grafting and is still used by orchardists when changing a fruit tree over to a different cultivar. It is relatively easy on medium to large limbs with a thickened bark layer such as apples and pears, although it can damage the wound area especially if the wound area does not heal completely after grafting or if it becomes diseased. It is

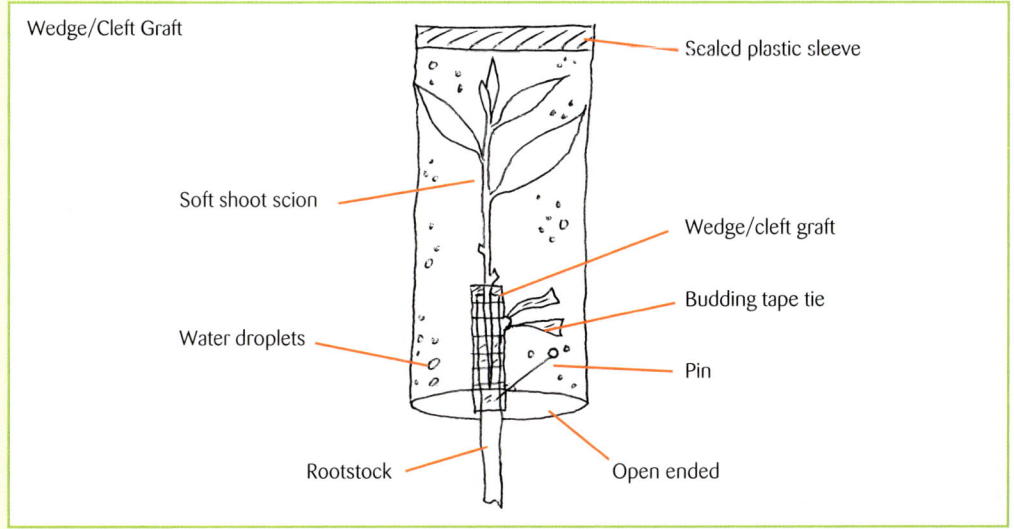

Wedge/Cleft Graft
- Sealed plastic sleeve
- Soft shoot scion
- Wedge/cleft graft
- Budding tape tie
- Water droplets
- Pin
- Rootstock
- Open ended

not recommended for large limbs of citrus because of their thin bark and the risk of damage or infection, but can be used for small limbs up to 1cm in diameter.

The tree or limbs are cut neatly and cleanly and scions inserted into the cut limb ends. An axe or pruning knife or special cleft graft blade and mallet can be used to make the cut. The implement used depends on the diameter of the branch to be grafted.

This graft method differs from the bark graft in that the whole limb is split through its centre from the cut end downwards and held open whilst scions are positioned at the bark edge; the split gap is then allowed to close, clamping scions in position. The cutting implement used to make the wedge cut can be used to hold the cut open until scions are inserted.

Scions are prepared by cutting both sides of the scion into a 'V' with gentle sloping edges. If a limb with a thickened bark layer is grafted using small scions, it is very important that the cambium layer on one side of the scion line up perfectly with the cambium layer on the rootstock otherwise the scions will not graft. The scion should line up with the cambium layers inside the outer bark layer. This is so that the cambium layers (just under the bark layer, where healing tissue is situated) will line up; at least one side of the scion must be aligned with the cambium layer of the rootstock. The grafted area and all cut surfaces of the limb should be protected with grafting wax after inserting the scions. Small grafted limbs can be wrapped with budding tape and tied securely. Again the use of plastic sleeves is recommended.

Soft-shoot cleft graft

The cleft graft is also used for citrus when grafting soft growth shoots (scions) to soft growth rootstocks when citrus are grown in pots in a protected environment and when *in vitro* grafting is done for virus indexing. This method is not recommended for home gardeners because of the degree of difficulty. The soft shoot cleft graft is delicate and has been used for inserting scions into just-germinated seedlings. Thin knives or razor blades are needed to do this work. The soft shoot scions can be tied with soft, narrow plastic ties, held with grafting pegs or pins, and grafts should be covered with plastic sleeves to ensure success.

Whip-and-tongue graft

This is the most common type of graft performed by home gardeners as it is fairly simple and can support itself effectively when done. It is only used for citrus when scion and rootstock pieces are of approximately the same diameter. With this graft both the scion and the rootstock are cut at an angle of about 30° to create a sloping cut through the stem or lateral. An extended cut at a slightly lesser angle is sometimes done to ensure better cambium contact and longer cuts are also made to support large scions. If the sloping cut is made at a more acute angle the pieces will not fit together very well and may fail to graft properly. After making the 30° sloping cut, another cut is made at right angles across the sloping cut surface and the knife edge slipped 2–3mm into the wood then along almost parallel to the first cut. This creates the 'tongue' of the whip-and-tongue graft on both the scion and the rootstock.

One of these cuts should be made about one-third the way along a sloping cut on the rootstock or the scion and the other begun about half way along the sloping cut on the other piece; if the two cuts are made at identical places the graft will not 'fit' together nearly as well. If the scion and rootstock are the same diameter then, when the graft is completed, it will be difficult to even see where the pieces are joined.

The top of the cut surfaces of the scion and rootstock are placed together and a gentle pressure applied to push the

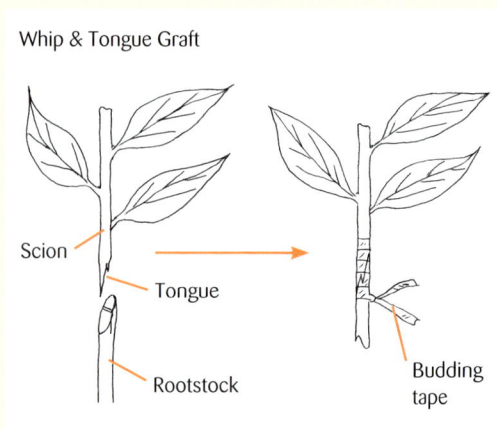

Whip & Tongue Graft

Leaves can be cut in half to reduce transpiration, or removed and scion enclosed in plastic film wrap. A plastic sleeve is then placed over graft and a paper bag used to cover and shade the graft union.

tongue pieces together; if the cut pieces overlap when pushed together more bark will have to be removed from one piece or the other or a piece of material removed so that the cut surfaces match. Some grafters cut the tip of each sloping cut so as to allow a 'half moon' cut surface to show each side of the graft, the theory being that the wound area will form callus more readily and thus form a very strong graft union. If the scion and rootstock pieces are of uneven diameter then the whip-and-tongue is adjusted in that the scion is shifted to one side, matching bark sections together to align the cambium layers. The other side of the graft union will have a large wound area; this can be left as is, or using a very sharp knife the open wound area can be pared back to the scion to make it rounded, although the total wound area is similar it allows the callus tissue an easier surface on which to grow and heal. I have used this method with various sized scion and rootstock pieces and the paring seems to work very well.

VIRUSES THAT CAN CAUSE PROBLEMS IN CITRUS WITH BUDDING AND GRAFTING

Exocortis (Scaly Butt) Scaly Butt is often seen on grafted trees using Citrange, Rangpur lime and Trifoliate orange rootstocks. It causes a scaly or peeling bark to develop below the bud union site. Some rootstocks, however, can carry the virus without showing visible effects. The virus is transmitted by grafting, including accidentally by grafting tools that are not cleaned and are in constant use. Clean equipment and virus free scions or buds reduce infection rates.

Tristeza (Quick Decline, Stem Pitting) This virus is endemic to Australia and causes a high percentage of citrus tree deaths. More virulent strains cause sudden tree death. Less severe symptoms include stem pitting, fruit malformations, and unthriftiness, but some infected material will show no symptoms at all. The virus can be transmitted by the black citrus aphid insects and by budding and grafting with infected material. Usually, all citrus plant material (in Australia) has a form of Tristeza. Fortunately, most commercial rootstocks are resistant but sour orange is an exception and when used as a rootstock, the tree can show quick decline. As lemons show more tolerance to this disease, sour orange rootstock tends to be used only with lemons.

Psorosis This condition relates to a group of viruses that cause symptoms such as crinkle leaf and, most commonly, the colouring of immature leaves from dark green to yellow-green patterns, not unlike some nutritional deficiency symptoms. The virus is bud transmitted so the use of virus free budwood will prevent infection.

Xyloporosis This disease causes poor growth, stunting and gradual tree decline, especially on cumquats, mandarins, tangelos and sweet limes. Avoid it by using virus free buds.

CHAPTER 4

managing citrus trees

Traditionally, Australian gardens have incorporated a lemon tree or cumquat. There are now many more citrus species available to the home gardener so selection will depend upon personal preference, location, the micro-climate and growing conditions. Before choosing a tree have a look to see if other gardeners are growing the same cultivar in your immediate area.

Citrus trees need high-nitrogen fertilisers to promote constant growth and adequate water is a major issue even during the winter period. Care of the tree involves controlling any pest or disease attack and pruning where necessary. Citrus trees can also be maintained in pots but will need regular fertiliser applications and regular repotting. In cooler areas, it is easier to provide wind and frost sheltered positions for potted or espaliered plants especially in small garden areas.

This chapter deals with some of the most important aspects of planting and maintaining citrus in home gardens, including maintenance of citrus in pots and shifting and repotting citrus trees. The focus is on organic fertilisation and management using low impact approaches.

Open-grown citrus trees

Planting out

If buying citrus trees for planting out, it is best to choose healthy trees with lots of dark green foliage and of good size rather than small and spindly. Avoid trees showing any sign of pests or diseases, especially citrus gall wasp or citrus leaf miner (see Chapter 6).

Most citrus trees available from plant nurseries are grown in pots and are available all year round, but for best results the tree should be bought in spring. Trees grown for a short time in pots will have a well-defined

root ball. This can be teased gently as it is repotted or planted out. Any suckers or roots growing in the wrong direction or coiled inside the pot should be removed before planting. The procedure is the same for home-grown or grafted citrus trees.

Because of the range of climatic conditions in Australia, careful consideration needs to be given to the location of citrus trees. The site chosen for a citrus tree should receive maximum sunlight (at least 5 hours sunlight per day) and be protected from strong winds and frosts. In tropical and semi-tropical areas, protection from cyclonic winds may be necessary and in some hot, dry areas, some shading from excessive heat with shade-cloth material may be required.

For planting trees, dig a hole about the same size of the root ball to plant into. The only exception to this rule is in heavy or tight (compacted) soils where about a one-metre square is dug out to loosen the surrounding area.

A small handful of organic fertiliser such as Dynamic Lifter™ can be placed at the bottom of the hole covered with some soil (between 5–10cm) before placing the tree into the hole. The tree roots should not come into contact with the fertiliser as it will burn the roots and prevent early establishment.

Young citrus tree just planted and staked with chives as companion plants planted around the tree base

Frost damage to 'Lisbon' lemons

Do not place the trunk any deeper than it was in the pot as this invites root rot or collar rot to infect the trunk.

Position the tree then lightly fill the hole around the tree roots. Tamp the soil but not too hard, then water the area, allowing it to drain; water it again with a seaweed solution (using one of the seaweed products such as Maxicrop™) to promote new roots. No extra solid, liquid or granular fertiliser should be given to the tree until it produces some new shoots. Then, it can be given a complete fertiliser or one of the organic fertilisers at a rate of about one handful per tree. Fertilisation at this rate can be carried out every month or two until the plant stops growing new shoots.

When planting into 'heavy' soils (soils with a large clay component) it is wise to build a large mound of soil to plant into; this will allow some of the shallow surface roots to breathe if the area becomes waterlogged.

A large (weed-free) tree guard around young trees or a trunk guard will prevent ring-barking by animals and give some protection from wind and frost in frost-prone areas. Tree protection too close to the trunk may encourage a build-up of pests such as snails, so check regularly. As an alternative, painting the tree trunk with thick white paint reflects heat without providing a haven for insects.

Fertilisers

Fertilisers for citrus trees usually contain plenty of nitrogen (N) and both phosphorus (P) and potash (K) at a ratio of about 13N:3P:3K. In the wet tropics where nutrient leaching is a major problem, the mix is more likely to be 12N:14P:10K. The NPK ratio is usually marked on the fertiliser bag but not necessarily for organic manures. In general, nitrogen is needed for leaf shoot and root growth and helps with fruit colour and size; phosphorus is needed for flower and fruit development and quality fruit production; and potassium improves fruit quality and tree health. As well, most micronutrients (including zinc, magnesium, iron, manganese, calcium and sulphur) are necessary for proper growth and maximum cropping. Too much of even one element, however, may give unwanted symptoms such as thickening skins (nitrogen) and delayed fruit maturity: for instance, an oversupply of nitrogen and potassium fertilisers will increase fruit acidity and delay fruit maturity. It is important, therefore, that fertiliser applications are balanced.

Once a young tree has started to grow vigorously (after the first year or two) it can be fertilised by gradually increasing the amount given from about 30g to about 0.5kg. A tree of about 10 years of age will require between 1.5kg–3.5kg of artificial fertiliser or 20–50kg of well-composted animal manure, while larger trees need more than this to ensure maximum cropping rate. Commercial growers use nitrogen (in the form of ammonium nitrate or ammonium sulphate or urea), superphosphate (source of phosphorus) and potassium sulphate (as a source of potassium) and apply these separately, the superphosphate and potassium being applied in bands around the 'drip line' of the tree.

Fertiliser application timing depends on the area where the trees are grown and when growth flushes occur. Ideally, nitrogen should be available for each growth phase and applied just before the expected growth flush. Trees can be given a balanced fertiliser in autumn, and a high nitrogen fertiliser to coincide with expected growth flushes (February–May, July–August, November–December). Organic gardeners may simply give regular applications of organic fertilisers such as chicken manure during the growth period from spring to autumn.

Organic or chemical fertilisers should only be applied to damp soil to prevent fertiliser burn; as an alternative the fertiliser can be watered into the soil after application.

Organic fertilisers are preferable and can consist of a regular application of animal manure and compost and urea for liquid nitrogen. Fertiliser should be applied to the ground to cover the area from the trunk out to about one metre past the edge of the reach of the tree's foliage, making sure the fertiliser is not touching the bark on the trunk. Commercial growers and some gardeners apply 'bands' of fertiliser around the 'drip line' of the tree. I once worked on an orchard where there was plenty of well-rotted poultry manure; a thick 5-cm mulch of this was applied around the tree to one metre beyond the foliage area during late summer immediately after the area was weeded. A follow-up application of a high nitrogen fertiliser was applied in spring. If fresh poultry manure without straw or trash in it was being used, the layer would have to be much thinner.

Organic manure may need a 'top up' with extra potassium and phosphorus fertiliser; this can be provided by rock dust or pelletised organic fertilisers containing the necessary nutrients. Some gardeners use blood and bone as a basis for citrus tree nutrition but because it contains only nitrogen and phosphorus, it needs to be supplemented by adding a small handful of sulphate of potash to each kilogram of blood and bone; a very small amount of trace elements may also be added. Organic manures have been shown to aid the health of citrus trees by preventing some diseases and pests from becoming a problem (see Resources section).

On acid soils or where acidic fertilisers such as sulphate of ammonia have been used constantly, it is essential to apply lime around the tree to reduce pH, and to supply calcium. On highly alkaline soils gypsum can be used to supply the calcium needed. (pH is a measure of acidity or alkalinity. See Glossary.)

Very often citrus trees show nutrient deficiencies (see Chapter 6); common among these are boron, magnesium, manganese, zinc and copper deficiencies, and it may be advantageous to spray liquid foliar fertilisers onto the foliage to correct these symptoms. Before applying foliar nutrients check the soil for acidity or alkalinity as these or conditions such as waterlogging or water stress may be causing the deficiency symptoms to occur and foliar nutrients may not be needed.

A few foliar fertiliser recipes are:

Boron foliar spray Mix 30g of borax with 10 litres of water. You will probably only need to spray once every 10 years.

Copper foliar spray Apply normal copper sprays (e.g. Bordeaux) as recommended for fungal disease control and this will supply enough copper to correct any deficiency.

Magnesium foliar spray: Mix 150g of magnesium nitrate in 10 litres of water to which is added a wetting agent. This spray should be applied when needed just after full flowering has finished.

Manganese foliar spray Mix 10g of manganese sulphate with 10 litres of water and apply when necessary to the spring growth.

Zinc foliar spray Mix 10g of zinc sulphate with 10 litres of water and spray onto established new spring growth shoots about November in years when needed.

One application of a specific foliar nutrient will usually be enough to last one full season. Do not apply foliar sprays in very hot, dry or sunny conditions as burning of the leaves can occur. Use a separate sprayer for applying foliar fertilisers but do not use one that has previously been used for chemical products such as weed killers. Apply one nutrient spray at a time because chemical reaction can occur between some chemicals.

Application of a seaweed foliar spray will also help the tree to take up some of the minor nutrients that it needs. Organic citrus growers use liquid seaweed such as Maxicrop™ combined with separate applications of a fish emulsion such as Vitec™, or a combined seaweed and fish emulsions such as Powerfeed™ as alternatives to chemical foliar sprays (see Chapter 6). Sprays are usually applied just before a growth flush begins, after flowering or just as the new leaves are expanding. Although spraying foliage with nutrients is called foliar spraying, the nutrients applied can be absorbed through the trunk, bark and twigs as well as through leaf surfaces.

Fertigation

Many citrus growers are now using micro-irrigation techniques to water citrus trees, such as drippers and mini fan sprays, or basal flooding – watering upward from the base of a plant (when potted citrus are grown in igloos), while at the same time providing all necessary fertilisers as liquid nutrients through the irrigation system. In this way irrigation and fertiliser application occur at the same time. This process is called fertigation. This method is expensive to install but very effective in weed control, nutrient uptake, reduction of water use and savings in labour costs, and it is available for home gardeners.

Watering citrus trees

The importance of watering citrus trees, particularly young trees, has already been referred to. Citrus trees require good moisture supplies because they are mostly shallow rooted with fine fibrous roots prone to rapid drying out and dehydration. Young trees need water to form a spreading root system and older trees need moisture at the time of flowering and fruit growth and for high quality juice to be formed in the fruit.

The need for water will vary with tree age, soil type, soil management and the micro-climate around the tree. Hot dry winds will, for instance, dry the tree and soil out very quickly. It is important to keep the water up to the tree during hot weather conditions, when a good soaking may be necessary during very hot spells or every day or twice a day for pot grown plants.

One of the mini fan irrigation systems now available

During cooler periods, watering may be necessary only once every 4–5 weeks or less. Even trees grown in a warm place during the winter can become dry through lack of water. Regular watering is very necessary to create healthy, green foliaged citrus trees. I have seen a lemon tree growing only a few leaves every year but nothing lush. The tree 'picked up' immediately the owner began watering during the cool winter period as well as in summer. The lemon crop improved and the tree produced much more foliage and fruit during the following year.

Potted plants will need watering every day in summer and very infrequently in winter.

Citrus trees cannot tolerate over watering so be careful; test for soil moisture with your fingers or use a moisture probe before watering. With pot-grown plants check that the drainage holes have not been blocked with root growth.

Irrigation systems available for home gardens range from sophisticated micro or trickle irrigation systems to perforated or soaker hoses to watering with a hand-held hose or flood irrigation. Because water restrictions are becoming more and more frequent, a micro-irrigation system, turned on only when it is needed, may be a good choice. There are several types including dripper systems and miniature fan spray systems. Beware of fully automated micro-irrigation systems as they can deliver too much water at the wrong times unless used with soil sensors to monitor soil moisture content.

Potted citrus tree showing effects of too much water

Mulching citrus

Mulching is an important aspect of citrus tree management. A thick layer of mulch material such as pea or wheat straw around the tree during the summer will help conserve moisture and reduce the need to water. Mulching trees in areas that have wet winters may be detrimental to tree health as it can increase the potential for waterlogging.

Do not place any mulch directly against trunks of citrus trees as this encourages root and collar rot.

Weeds growing near trees need to be kept under control as they use water and can contribute to the build-up of diseases and some insect pests. Weeds can be turned to advantage as mown mulch (remembering to leave half a metre around the tree trunk clear of mulch). Grasses and weeds or cover crops (such as legumes) can be mulch-mown or cut with a whippersnipper, gradually improving the soil and encouraging biological activity.

Shifting citrus trees

Citrus trees adapt very well to shifting with good preparation and after-care to allow maximum recovery.

The best time to shift citrus trees is when the weather is warm but not hot enough to cause heat stress. It should not be too cold either. This means that early spring and late summer are the best times.

I have shifted many citrus trees using the technique detailed below for shifting 'in ground' citrus trees from one place to another.

1. Dig and carefully prepare a hole at the new site ensuring good drainage. The hole should be deep and wide enough for the root ball of the tree being shifted.
2. On the tree to be shifted, first cut back the foliage to compensate for the loss of roots that will be necessary when the tree is dug out. The foliage area should be reduced by at least one-third, but by as much as two-thirds for an unhealthy tree or one with a very large canopy and a small root ball. Do not cut back each branch by two-thirds; rather, remove a few limbs totally. This will keep the original tree shape, provide shading and assist recovery after shifting, especially

Transplanting lemon tree – root ball being enclosed to enable shifting

Surplus fruit removed from the tree can be used for juice, or dried fruit delights such as chocolate-covered orange slices or for preparing candied fruit or skins. The skins of sour orange, lemons, grapefruit and pomelo can be used for preparing marmalade jam. Cumquat fruit can be kept and preserved by making brandied cumquats.

during hot weather conditions. It is advisable to chamfer (see Glossary) the end of the remaining branch stubs, creating a sloped, cut edge to bark level to allow easier callusing to occur; also paint the cut edge with wound dressing paint (see Glossary and Chapter 5 on pruning).

3. Remove any fruit to reduce stress on the tree.
4. The aim now is to form a root ball, able to be lifted, but at the same time minimising root disturbance. All the earth should remain around the roots within the ball. The size of the root ball depends on the size of the tree to be shifted and the equipment available for lifting and moving the tree to its new position. In general, the larger the root ball the better it is for plant recovery. As a guide, large trees should have a root ball of approximately 2m, but if heavy machinery (e.g. bobcat or backhoe) is available, the root ball can be larger. To begin, dig a circular trench around the tree at a distance determined to be the outer ring of the root ball. A trench about 50cm deep should be enough to cut most of the roots, as citrus are surface-rooted plants. After digging the trench, define the base of the root ball by digging underneath the tree from the side, cutting any roots that are found. In deep well-drained soils, some roots will have grown downwards and these can be hard to get at and cut. Sometimes it is easier to dig one side of the root ball then lean the tree to the dug side to reach the other half of the root ball base. The result of this step should be a distinct 'ball' of roots enclosed by soil.
5. Trim the exposed and roughly cut edges of roots and chamfer large roots and paint with wound dressing paint. (Do not use oil-based paint, as this will poison the tree! See Glossary.)
6. Once the root ball is free, wrap it in hessian or, alternatively, slide an old piece of carpet or matting underneath it so that it can be dragged to the new site or loaded onto a truck lifted by machine.
7. Place the tree in its new hole so that none of the trunk is buried when replanted. Fill around the edges with soil and gently tamp the soil into place. Do not ram soil as this will inhibit water penetration around the root zone. Once the hole has been filled, water the area thoroughly. After a few hours water again, this time using a liquid seaweed mix to promote new fibrous root growth.
8. The tree may need to be staked until new roots form to anchor it properly. Staking will prevent damage from strong winds that may

wrench the tree. Use two or three good solid stakes and tie with flexible ties (e.g. strips of inner tyre tubing) in a figure-8 configuration around the stakes with an extra loop around the plant stem.

9. It may be necessary, depending on weather conditions, to put some shade-cloth covering over the tree or on the sunny side for one summer period to prevent sunburn to leaves and bark. If some leaves fall and too much bark is exposed to sunlight then paint limbs over 3cm wide with whitewash (a powdered lime and water mix) or a water-based white paint to protect from sunburn. Do not use oil-based paints!
10. To enhance recovery, the foliage area can be sprayed once every three weeks or so with a very weak solution of liquid seaweed such as Maxicrop™ to help the tree recover quickly. Seaweed products also provide some micro nutrients that are absorbed through the leaves. Once the tree starts to produce new growth shoots then it can be given small amounts of a complete fertiliser every month during the growth period to further enhance recovery. A large lemon tree I transplanted was fully recovered and fruiting again within eleven months of shifting.

Large citrus trees can be lopped to virtual stumps before shifting. Follow the procedure in Chapter 5 for pruning to a stump, and then follow the steps for shifting given above and the tree should recover fully. Shading the tree with shade-cloth until new shoots grow from the stump is also a good idea.

Citrus in pots

Citrus plants in large pots are ideal for modern small gardens as they can be placed anywhere that receives plenty of sunlight and shelter from winds and frost: on balconies, in small courtyards, or on rooftops. Flowering citrus gives off a heady perfume and citrus trees in pots with full crops of brightly coloured fruit look spectacular.

Most citrus trees will grow well in pots, although cumquats, calamondins, lemons, mandarins, lemonade trees and sweet oranges are good choices where space is limited.

Citrus trees can also be grown successfully as bonsai plants but care and maintenance for bonsai plants is much more demanding and will not be covered in detail here.

Potted citrus need regular fertilising, but it is important to apply a little often from spring to autumn for best results. Too much fertiliser will cause salt burn, make the soil become too acidic or cause signs of nutrient deficiencies to appear on leaves. Regular watering is also essential to maintain tree health and vigour as potted plants can dry out quickly even during cool dry weather periods. A pot in a very hot position can have a wrapping of aluminium foil placed around it during the mid-summer period to prevent heat stress.

Young pot-grown citrus tree showing upward growing roots and congested root ball

'Lisbon' lemon in pot growing in Hobart, Tasmania

Repotting citrus trees

Many gardeners wait for too long before repotting citrus, resulting in trees becoming root bound, losing vigour, dropping leaves and becoming unthrifty (unhealthy) with yellowing and sparse foliage. A citrus tree can remain in a pot without repotting for 3–5 years depending on the size of the tree and the pot and the growth rate of the tree. When young trees are placed in small pots they will outgrow the root area within the pot more quickly than older trees in larger pots.

There are two relatively easy ways to repot citrus. One is to completely remove the plant from its container and repot it. The second is a method I have developed which involves leaving the plant in its pot and repotting in three stages. The second method is particularly good for plants that are too large to remove from their pots.

The first method involves reducing the foliage bulk by pruning to a reasonable shape, gently pulling the plant from the pot and laying it on its side. Using a hand saw or old bread knife, remove the outer layer of massed roots to 3–5cm in from the outer edge. This will reduce the total size of the root ball. Place the tree back into the same pot or a new one with some new potting mix already in the base. Pour a good free-draining potting mix around the root ball to fill the pot. Use a short narrow stick to help push the soil mix into the cavity between the root ball and the side of the pot to ensure there are no large air gaps. Make sure to leave enough room at the top of the pot to allow a watering trough; if the pot is overfilled water will not be able to penetrate to the plant roots.

Repotting citrus—hollow plastic pipe used to drill holes for plant when repotting

Water the repotted tree thoroughly, leave to drain for about an hour, then water again with a liquid seaweed solution (made up using concentrated liquid seaweed extract) to encourage new fibrous root growth. Place the repotted citrus in semi-shade for a few weeks (or cover with shade material) to allow it to recover before placing the pot in full sunlight.

Once new shoots appear, make sure the tree has a regular supply of nutrients by giving it small amounts of pelletised fowl manure or a slow-release fertiliser every month whilst the tree is showing new growth shoots. Monthly applications of a very weak liquid seaweed solution as a root drench and foliage spray will enhance tree recovery and protect against some insects and diseases.

In the second method, rather than taking the tree out of the pot, sections of the roots and accompanying soil are treated *in situ*. This

Repotted and pruned cumquat tree used as an ornamental garden feature

involves segmenting the root area and soil in the pot into nine wedges and removing these one at a time over a period of three years.

Three segments are removed in the first year, three different ones the next, and three different ones the next, by which time all the soil and roots will have been totally replaced. The best time to do this is in spring or when the weather is warm. The technique is as follows:

Year 1 Using powder or spray-on paint, mark three wedges on the surface of the soil in the pot. Mark the edge of the pot itself as well to identify where the original segments were aligned for years two and three. Divide each wedge into three segments as per the following diagram.

David Goldsworthy removing outer roots from original root ball before repotting

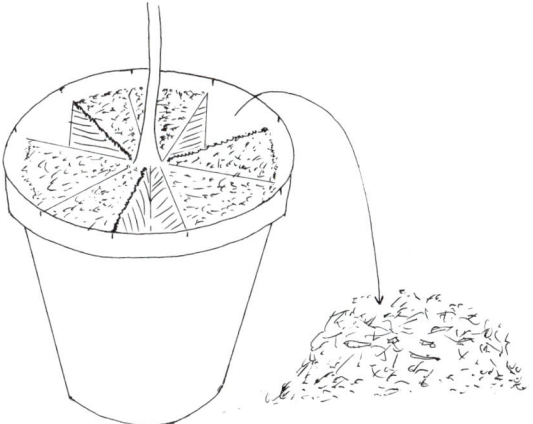

Remove three segments equidistant from one another, including all soil and roots, and fill the resultant spaces with new potting mix. Drench the whole root system with water, then a little later with liquid seaweed solution. Slow-release fertiliser can be given to the plant two weeks after the operation.

Year 2 Repeat the operation, removing marked wedges next to the ones removed in year 1.

Year 3 The remaining three wedges can be removed following the same procedure as in previous years. This will remove the last of the original soil and root mass.

There should be no need to repot the tree for another three to five years, depending on the health of the tree: repotting will be unnecessary if the tree is healthy with plenty of green leaves and good fruit.

An organic citrus grove

Robert Ridgewell and Jan Denham live in Palinyewah, NSW, an isolated area not far from Mildura near the banks of the Darling River. They own and operate a 21ha property, 'Karra Organic Farm', growing a range of citrus trees, plum trees and a few vegetables using organic farming and orcharding techniques. Jan has been very active and one of the leaders in the organic movement for many years. Robert and Jan are fortunate to have Robert's brother Ian Ridgewell working with them. Although not in partnership, his 25 years' experience on the property is invaluable.

The property has been managed organically since 1989. Citrus cultivars grown include Imperial mandarins, Leng Navel, Navelina, Washington Navel, Lane Navel, Valencia, and Tangelo. Plum cultivars grown as a supplementary crop include 'Tegan Blue', 'Autumn Giant', 'Ruby Blood Plum', 'Radiance' and 'Roysum'. Approximately 90% of the fruit produced is marketed locally within Australia and a small amount sent to overseas markets.

'Valencia' orange grove with weeds controlled around trees and cover crop in centre of row

As with most organic farming operations, Karra Organic Farm relies upon feeding the soil and improving the organic content within the soil profile to increase productivity. The soil on this farm is very sandy, which is ideal for growing citrus, but unfortunately it is also very impoverished so needs constant feeding. When Jan and Robert began, the organic content of the soil was less than 0.3% but with care, good management, the use of cover crops and compost addition, the organic content of the soil is now a respectable 2 % and rising.

Compost is made on site and composed of fowl manure (cow manure will be used as a replacement from 2006 onwards to reduce high phosphate levels), sawdust and grape marc, and is able to be prepared in just 8–10 weeks. The compost is frequently turned and water is added as necessary so the hot compost mix reaches a temperature of 60°C, a temperature needed for proper management of disease organisms and weed seed eradication. Robert says that his dog once found a rabbit that had burrowed into the heap, by then nicely steamed. Robert also says that the compost gets hot enough to cook potatoes wrapped in foil.

Compost is usually applied to citrus trees in winter as a thick band 20cm wide around the foliage drip line area of the trees.

Other nutrients for the citrus trees come from cover crops such as grass and lucerne grown between trees and regularly mulch-mown. Barna grass (a species with sugar cane like growth) has also been used for windbreaks and for mulching material that is ploughed into the soil or used as a surface mulch. Leaf analysis is regularly done to indicate nutrient needs of citrus trees and several foliar sprays are applied during the year using a nutrient mix that includes about 14 different materials including kelp seaweed (dissolvable crystal form), to correct the mineral imbalance, sulphates of various minerals including zinc, magnesium, manganese, and potassium are applied as foliar sprays. Tiny amounts of cobalt and molybdenum are also applied to trees. To build up the biological activity within the soil, fulvic acid and fish extracts are applied through the dripper system and micronised calcium and gypsum are spread around the trees regularly.

Copper foliar sprays are used as needed to provide nutrient copper to the trees but the necessity for regular copper sprays to control diseases has lapsed because there are now minimal fungal diseases present in the citrus grove. It is interesting to note, too, that many copper sprays can kill some of the soil-inhabiting organisms thus indirectly affecting soil structure. In the past there were several pests or diseases attacking trees, foliage and fruit but since the early 1990s no pesticides such as oil sprays for scale control have been used. The insects occasionally seen include Katydids, a grasshopper-like insect, and the Spined citrus bug, but as both of these cause little damage they can generally be manually removed from trees. Very few mealy bugs (a major pest of citrus) are found, but on nearby citrus properties where chemical control measures are constantly used this insect is a major problem as are scale insects. Predators such as wasps and ladybird beetles are present in large numbers at Karra Organic Farm as are birds and all these help control insects. Diseases such as sooty mould are not seen because the large population of insects needed to excrete 'honeydew' (a sugary liquid) that the mould lives on, are not present. Citrus leaf miner has not been a problem in the past few years probably due to the cold winter periods and the build-up of insect predators within the citrus grove.

Orange half skin with fungal growth as the skin and pith breaks down during composting

Phytophthora, a root disease, has been a problem on mandarin trees grafted to sweet orange rootstock but this was worsened by very saline water supplied from the Murray River during 2003, weakening the trees making them susceptible to the disease so that some were lost. Replacement plantings have begun and the aim is to remove all citrus including some older weaker trees that are on sweet orange rootstock. Replacement trees are grafted onto citrange or Trifoliata rootstocks that are more salt-tolerant and more resistant to nematodes. All citrus trees are skirted to allow airflow into trees and prevent slug and snail access to foliage; old dead twigs and branches are pruned from any weakened trees, and crossing or strong centre branches pruned off.

There are no significant pest or disease problems affecting the plum trees and this is due in part to the excellent climate producing very dry weather during the spring and summer period.

Vegetables are grown on the property, some between citrus trees, and they include cabbages, zucchini and butternut pumpkins. Pests and diseases on these are minimal with the cabbage moth problem controlled with Dipel™ biological sprays at 10–14 day intervals. Milk and liquid compost tea have been used for powdery mildew of pumpkins but as the disease is slow to develop, spraying is not essential. Any abnormally growing or seemingly diseased plants are pulled out and destroyed to prevent disease build-up. During 2005, earwigs became a problem by eating tender young pumpkin plant foliage.

Regular hand thinning is used to overcome alternate cropping and the production of small citrus fruit. The thinning standard is to allow 7 fruit per half metre cubed of foliage area. Production from Karra Organic Farm is up to the district average: for instance Valencia oranges produce from 30–70 tonnes per hectare (allowing for alternate bearing years). Fruit is picked at optimum maturity.

Mandarins are picked with cotton-gloved hands to avoid injury to the skin and once picked the fruit is 'dry-brushed' to remove dirt and to polish fruit. The waxy shine from dry brushing the fruit remains for 3–4 weeks. Fruit is packed for delivery with the

Organic citrus grove, Robert, Damian and Ian Ridgewell – mandarin picking done with cotton-gloved hands

second quality fruits placed in netted bags and best quality fruit in boxes. Fruit is cool stored at 6°C.

Irrigation in the past involved overhead spray irrigation then low profile fan irrigation techniques but now drip irrigation is the only method used. The old overhead spray irrigation equipment is still in place and available to be used for frost control when necessary. Soil moisture probes are used to determine the correct time for irrigation. Fertigation (irrigation water containing soluble fertilisers) has been used to water and feed cabbages.

Commercial citrus growers obtain rootstock seed and virus-tested buds for propagation purposes from Auscitrus (see Glossary) to ensure clean propagation material. Home gardeners too can buy seed and budwood from Auscitrus; the only proviso is that a minimum of a $50 order is forwarded. If this amount of money is too much then gardeners can make a bulk order through garden clubs or other organisations so that the cost is shared. The postal address for Auscitrus is PO Box 267, Dareton, NSW 2717. Information about Auscitrus and on the material available (seed and budwood) can be viewed on Auscitrus website: www.auscitrus.com.au

CHAPTER 5

pruning and training citrus

George Quinn, a Horticultural Instructor in South Australia, recommended as long ago as 1921, early pruning of citrus from the time of planting and for several years thereafter, to shape trees and to prevent intertwining branches from forming. His comments on pruning orange and lemon trees and the treatment of fruiting wood are still valuable today. He makes the point that the best fruits are borne inside the body of lemon trees, while the best oranges are found on the outer tree canopy. The terminal twigs alone of orange trees produce first-class fruit, while on lemon trees the laterals (see Glossary) all along the horizontal side limbs will bear.[23]

Quinn also assesses the value of side shoots. Strong upright limbs or watershoots (see Glossary) absorb most sap and possess most vigour but produce no fruit for some time, while less stimulated or vigorous shoots will bloom. In pruning light wood on the branches, vertical shoots are suppressed and those on the sides retained. Thus, he argues, the key to pruning lemon trees for fruit production is to check (prune) vertical shoots and only encourage the horizontal ones. The sides of the tree should be built up from the bottom; the centre will always provide necessary branches to raise the top when required. As laterals extend away and hang down with fruit, they should be shortened back towards the parent branch. Quinn argues that no crowding should be permitted, and crossing or tangled branches as well as dead or worn out branches have no place in a well-kept lemon tree. Such precautions prevent scratching and rubbing, a common cause of depreciation in lemon fruit.[24]

As Quinn argues for lemons and oranges, most citrus need new growth for fruiting, and pruning appropriately is one way of ensuring new growth in the following season for heavier fruiting. Correct nutrition and adequate watering will usually ensure new lateral growth. Some home gardeners regularly clip about one-third of new shoots from the growth flush that occurs on orange trees after harvest just to ensure larger oranges in the following season. Many home gardeners never prune their trees at all.

Generally, though, for healthy trees with vigorous growth, pruning can be kept to a minimum. Masanobu Fukuoka in his books on natural farming methods[25] gives a number of reasons for not pruning citrus at all, allowing trees to form a natural shape and size, with no tangled branches and no biennial bearing (see Glossary). There are, however, some situations where citrus trees do need to be pruned.

Pruning is necessary for training young trees, for espalier or topiary shapes, thinning fruit, (especially with some mandarin cultivars), preventing alternate (biennial) bearing (see Glossary) or for opening up and skirting the tree to allow air flow and sunlight into its centre for better pest and disease control. Citrus grown in pots or as bonsai also need regular pruning, as do hedgerows. Pruning is necessary when preparing trees for grafting or budding, shifting trees or resurrecting old or diseased trees (including 'skeletonising' trees to bring sickly trees back to health).

Pruning and training equipment

Cutting tools, ladders, ties and wound dressings are the major equipment of the pruner. Cutting tools and how to test them for sharpness have been dealt with in Chapter 3 on propagation. For pruning, secateurs, saws and pruning knives must be kept sharpened and clean at all times.

All cutting surfaces should be cleaned with a 10% methylated spirit–water solution or strong bleach, then rinsed with tap water. A rag soaked in the cleaning solution can be kept handy for wiping blades to remove grunge and sap. Cutting equipment must be sterilised immediately after pruning any diseased material before further pruning. Some diseases are easily transferred from one branch to another and can gradually cause the death of the tree. To store cutting equipment, wipe blades with grease or oil to inhibit rust and keep them in a dry place.

Ladders are necessary for pruning but they are a source of great concern to me as I have witnessed many rapid, catastrophic and unintended descents from poorly designed or inappropriately used ladders. Many of the available ladders are good on flat surfaces but inappropriate for sloping or uneven ground as might be found in gardens and orchards. Old, worn or bent ladders are a trap, especially if you are reaching for a particularly inaccessible branch. If you do a lot of pruning it is worth investing in a strong, well-designed ladder with wide legs, easy-to-clean rungs and efficient locking devices.

Tapes and ties are used for tying espaliered tree branches into place or to help secure trees to stakes. Budding tape is used for budding and grafting operations (see Chapter 3).

Wound dressing paints have been dealt with in Chapter 4 on budding and grafting. Their purpose is to seal a wound area to prevent bacterial and fungal infections.

Whitewash, a mix of powdered limestone, water and a few drops of oil (to help the paint stick) or water-based white paint are used to prevent sunburn after severe pruning and to protect the tree trunk from insect and disease attack. Oil-based paints must not be used as these will suffocate the bark and trees will die.

Skeletonising citrus trees

Pruning harshly to skeletonise a tree is done just before shifting a large tree (see Chapter 4), when a tree is infected with a disease or citrus gall wasp (see Chapter 6), when a sickly tree needs to be rejuvenated or after severe frost injury. Skeletonising pruning must be done when the danger of any frosts has passed, preferably during late spring or summer.

Remove all small twigs and small branches on the tree as well as any dead wood. This will leave the 'skeleton' of main branches bare except for a few leaves here and there. Once pruned, it may be necessary to spray with a product such as Pestoil™ or a copper spray to get rid of any scale insects or diseases (see Chapter 6).

All bark on the remaining branches and trunk must be painted with whitewash or non-oily white paint (see above).

The harsh pruning must also be accompanied by a good application of fertiliser to help produce new growth. Organic fertilisers such as fowl manure are recommended, although other fertilisers may be needed if the soil is deficient in particular elements (see Chapter 4). Apply fertiliser after pruning and after aerating the soil with a garden fork or small soil auger. Watering the root area with a liquid seaweed (such as Maxicrop™) soil drench will encourage new fibrous root formation.

Skeletonised citrus tree being rejuvenated by pruning. Trunk painted white to protect from sunburn

Skeletonised trees will soon produce healthy growth but may take up to two years to return to full size and normal cropping.

Pruning to a stump

Citrus trees are sometimes pruned back to a stump: the tree is cut down leaving only shortened pieces (stubs) of each major branch. This is often done when the tree is to be grafted, shifted or reshaped. It is essential to use a white non-oily paint (see above) on all the remaining trunk and branch stubs to prevent sunburn.

Pruning established trees

Older established trees require little pruning but home gardeners can prune every seven to ten years to keep a tree lower and more compact. An alternative is to cut out the centre of the tree leaving the other branches uncut. Dead twigs should be removed. Small branches with hardly any foliage can be lightly pruned to rejuvenate the tree.

Lightly pruning established citrus trees can improve the quality of the fruit and may indirectly improve cropping. Pruning too hard too often will, though, decrease cropping potential.

Some species of citrus, such as many mandarin cultivars, produce large crops of small fruit and have very dense foliage that inhibits light penetration into the canopy. These will benefit by pruning to give larger fruit size. The method is to remove some small branches within the tree canopy rather than prune shoots from all limbs, a method often referred to as 'chunk pruning'.

Cumquat trees grown in fruit fly infested areas can be pruned to create a late flowering and fruit setting regime so that the tree fruits when fruit flies are not active (in cooler weather).

Established citrus trees may have low branches touching the soil. These low branches can make a thick foliage barrier preventing air movement within the tree and providing conditions suitable for fungal diseases and access for such pests as slugs and snails. Low hanging branches can also stir up dust when blown by winds. 'Skirting', the removal of the lower foliage of the tree canopy, is sometimes done to deal with these problems.

Trees grown as hedgerows will need regular pruning every few years. Established trees in a hedge formation can be pruned every 3–4 years with hedge clippers or electric shears to maintain their shape.

A new method of pruning lemons

Home gardeners do not usually prune lemon trees until absolutely necessary because of disease or because the tree becomes too large. Often pruning entails just a light trim, which then has to be done every year because of the strong resultant regrowth. When a citrus tree becomes too large, it can be pruned so that the pruning job only needs to be done every 5–6 years. I have developed this approach for lemons, but it can also be applied to other citrus species. It must be noted that the 'Meyer' lemon does not usually need cutting back, as it tends to remain a small bushy tree.

To start this new pruning method, established trees are cut down to about one-third their original height, leaving long stubs of branches but retaining any other foliage below the cut ends. Neatly trim all branch ends. Wound dressing paint or grafting wax on the cut surfaces is optional but advisable to prevent cut limbs surface drying out.

After pruning, the tree will produce many new branches; these may grow one or two metres long in one year. These new, long and whippy branches should be left unpruned for 5–6 years. They will fruit heavily during their second or third season and the weight of the fruit will make the branches weep downwards. If the branches seem too stiff then attach weights to their ends to make them hang down. Kept in a bent position for a few months, these branches will remain bent, even after harvest of the fruit, resulting in a more mushroom shaped tree with a lower profile. Fruit are easier to pick and the tree is easier to manage. By contrast, if the tree is pruned harshly every year, or more often than suggested, long strong shoots will grow, shading the tree centre and reducing cropping.

Low branch of lemon tree that has been pruned, using the author's pruning method to create low hanging branches

Pruning young trees

Young citrus trees when purchased from plant nurseries can be loaded with fruit. Contrary to popular belief, it is not necessary to remove all these fruit before planting, nor is it necessary to prune the tree severely before planting out. The reasoning has been that the tree will be weakened severely by root disturbance at planting and will put all its energy into the fruit instead of new shoot growth.

If a tree is already fruiting and the tree roots are not disturbed during planting or repotting, my experience indicates that it is better to leave fruit on the tree and not prune at all. To compensate for the fruit load, the tree can be fed more frequently and watered well to prevent water stress. Watering the root zone with a liquid seaweed product will help produce new fibrous roots. If a young tree has an excessive fruit load (say, for instance, twenty fruits on six tiny branches) then it may be necessary to remove about one-third of these to make sure the remaining fruit grow to a reasonable size, but do not remove all fruit. Once a tree has begun fruiting it should remain fruitful from then on.

Young trees may have to be pruned to create a natural domed shape or set topiary shapes if desired or pruned very lightly for espalier shaping. No pruning should be necessary until between five and eight years after planting, when the tree is partially established and has a good supportive root system.

Topiary shapes

To create topiary shapes or bonsai shapes it may be necessary to prune several times every year to make sure the shape remains intact, but the best time to prune to obtain fresh new growth is in springtime.

Training citrus trees as espalier

Espaliered trees are trees trained to grow flat against a wall, on a frame or trellis. In cool areas, citrus trees will benefit from a warm protected spot and espalier trees trained against a warm wall protected from frosts usually perform well. In cold European countries citrus were grown in walled gardens for this reason. In tropical and semi-tropical areas of Australia, a frame or trellis in the open is a more appropriate location.

All citrus species can be formed into espalier shapes but the most commonly trained are lemon and lime trees. Lemon cultivars Lisbon, Meyer, Eureka, Villa Franca, the limes (Kaffir, Tahitian and West Indian), and the Lemonade tree are very suitable for growing as espalier shapes. Some of the newer Australian citrus cultivars will also be suitable for espalier growing. Mandarins and cumquats are often grown as topiary shapes but if grown as espaliers should be trained as flattened multi-branched fan shapes because of the thin spindly branches that are produced.

'Lisbon' lemon trained as an espalier

PRUNING AND TRAINING CITRUS

For citrus grown in pots a framework can be erected to create an espalier shape.

Espalier training can begin as soon as a young tree is planted out. If tree branches are growing outwardly in the wrong direction, they can be pruned off or they can be tied back to the framework. Selected branches can be tied down into the position necessary for the design shape and the foliage continually clipped to create a thin wall of leaves to maintain the espalier shape needed.

Espalier designs

There are many espalier designs to choose from but the most commonly chosen shapes for citrus are the multiple 'T', flat pyramid, open fan and thin rectangular or square hedge topiary forms.

Multiple 'T' shape For this shape, espalier branches on either side of the tree are bent horizontal then tied so as to space branches about 30–45cm apart. The topmost shoot is allowed to grow upwards as the next branch to be layered will grow from this. If the trunk is exposed to a lot of sun the trunk should be painted with a white water-based paint to prevent sunburn.

Open fan shape To create this shape, branches poking outwards from the flat plane needed are simply pruned off, and the remaining branches trained into a shape reminiscent of an open hand with the fingers spread wide.

Pyramid shape For this shape, the tree is allowed to grow into its natural shape, the only difference being that side branches growing to the front and back are removed to create a flattened espalier pyramid shape. The foliage is continually clipped to keep the shape in form.

Clipped hedge Several years ago some commercial growers started growing citrus in hedgerows, to save space and to make picking more efficient. Once the trees matured, they had to be kept in check. To do this, every second or third tree is removed, and once the remaining trees meet and form a hedge the foliage is kept under control using a bank of circular saws to prune the sides and the top of the hedgerow every 3–5 years. This opens up the foliage to light and allows fruit to develop all over the outside surface of the hedge. Hedges are often pruned with sides tapering in slightly towards the top to allow more sunlight to penetrate. Home gardeners can use this hedging technique to save space, stop foliage over-hanging pathways or to keep

Commercial citrus trees that have been pruned as hedges

trees at a reasonable height; because the major fruiting sites (90%) occur on the outer reaches of the canopy, selective tree shaping will not harm cropping potential. A hedge in a flatter plane can also be created using espalier training techniques; the hedge can be pruned wide or narrow (the thinnest hedge recommended is about 0.5m wide as citrus pruned too hard will just produce growth without fruiting), tapering or straight.

CHAPTER 6

pest and disease control of citrus

Citrus trees grown in Australia have many insect or animal pests, various diseases affecting the plant, stem, leaves or fruit, and a few disorders such as split fruit skins and nutritional deficiencies that show up readily when a mineral deficiency is present in the soil. 'What is wrong with my lemon tree?' is one of the most asked questions. This chapter attempts to help gardeners answer that question for lemon trees but also for citrus generally. First of all, to help home gardeners evaluate their own citrus tree problems and find a solution, the chapter begins with a list of symptoms. This quick guide to the likely problems is followed by comprehensive material on pests, diseases and nutrient problems of citrus together with information on organic control methods.

Quick guide to citrus problems

The following table provides a rough guide to likely problems. These can then be followed through in the later, more detailed alphabetical listing of citrus pests, diseases and disorders.

Valencia orange halved to show granulated dry flesh (pulp), see page 127

An example of one of the lichen species on bare branch of apple tree, see page 128

Lemon fruit cut in half to show the mutation of a fruit within a fruit, see page 128

Navel oranges showing very thick skins and granular flesh, see page 129

Citrus leaves showing effects of waterlogging on leaf colour, see page 130

Lisbon lemon stem showing multiple branching and over-abundant flower/fruit production that can be caused by nutritional deficiencies such as those of copper or sulphur

Quick guide to citrus problems

Leaf problems

Spots on leaves

These may be caused by
- citrus canker
- fertiliser injury
- fungal or bacterial diseases such as Anthracnose, Greasy spot (a *Diaporthe* spp., that causes oily wet-looking spots to develop on leaves and the leaf begins to curl), Black spot and Melanose
- insects such as scale, that can appear as spots on the leaves of citrus (see below)
- leaf blotch
- mites (see below)
- nutritional disorders
- salt burn
- sunburn

Discoloration

Colour of leaves can vary and be influenced by environmental conditions, excessive heat, water stress, diseases, viral infections or by nutrient deficiencies.
- Yellow leaves: cold, wet winters (cold prevents nutrient uptake and the constant moisture drains all nutrients away from the root zone of the plant).
- Yellow with dark veins or brown, water-soaked tips to the leaves: waterlogging preventing nutrient uptake can turn citrus leaves yellow.
- Brown and burnt leaf edges: highly saline water supply.
- Yellowish-white bleached colour: sunburn.
- Leaf-colour variations: high or low soil acidity or alkalinity can lead to nutrient deficiencies or toxicity.
- Black or blotchy-black leaves are usually covered with a mould fungus, sooty blotch (*Gloeodes pomigena*) or sooty mould (*Capnodium* spp.).
- Wilted black leaves: severe waterlogging or frost
- Discoloured leaves: Mites

Scale insects

- Black scale or Brown olive scale (*Saissetia oleae*)
- Californian red scale (*Aonidiella aurantii*)
- Chinese wax scale or Hard wax scale (*Ceroplastes sinensis*)
- Citricola scale (*Coccus pseudomagnoliarum*)
- Circular black scale or Florida red scale (*Chrysomphalus aonidum*)
- Cottony cushion scale (*Icerya purchasi*)
- Musel scale or purple scale (*Lepidosaphes beckii*)

Continued on next page

Quick guide to citrus problems

Scale insects continued	• Pink wax scale (*Ceroplastes rubens*); Soft brown scale (*Coccus hesperidum*); White louse scale or Citrus snow scale (*Unaspis citri*); White wax scale (*Ceroplastes destructor*) • Yellow scale (*Aonidiella citrina*).
Other insects or mites	Many species of insects can be found on citrus leaves or sucking or biting into them, the most common are: • Ants • Aphids • Bronze orange bug (*Musgraveia sulciventris*) • Citrus whitefly, Australian (particularly *Orchamoplatus citri*) • Compost flies (including Vinegar flies *Drosophila* spp.) • Crusader bugs (*Mictis profana*) • Earwigs, European (*Forficula auricularia*) • Mealybugs including the Citrus mealybug (*Planococcus citri*), the longtailed mealybug (*Pseudococcus longispinus*), and the Citrophilous mealybug (*Pseudococcus calceolariae*) • Mites: Citrus red mite (*Panonychus citri*), Citrus rust mite (*Phyllocoptruta oleivora*), Citrus bud mite (*Eriophyes sheldoni*), Broad mite (*Polyphagotarsonemus latus*) and the Brown citrus rust mite (*Tegolophus australis*) • Passion vine leaf hoppers (*Scolypopa australis*) • Thrips: the Greenhouse thrips (*Heliothrips haemorrhoidalis*), the Citrus rust thrips also called Orchid thrips (*Chaetanaphothrips orchidii*), and the Plague thrips (*Thrips imaginis*) • Weevils: Apple weevil (*Otiorhynchus cribricollis*), Citrus fruit weevil (*Neomerimnetes sobrinus*) the Citrus leaf-eating weevil (*Eutinophaea bicristata*), the Spinelegged citrus weevil or Dicky rice weevil (*Maleuterpes spinipes*), the Elephant weevil (*Orthorhinus cylindrirostris*), the Fruit tree root weevil (*Leptopius squalidus*), Fuller's rose weevil (*Asynonychus cervinus*) and the Whitestriped weevil (*Perperus lateralis*)
Holes in leaves	• Citrus butterfly larvae (caterpillars) of two species; the large Citrus butterfly (*Papilio aegeus*) and the small Citrus butterfly (*Papilio anactus*) • Grasshoppers, the most common being the Giant grasshopper (*Valanga irregularis*), the Australian plague lotus (*Chortoicetes terminifera*), the Spur-throated locust (*Austracris guttulosa*) and the wingless grasshopper (*Phaulacridium vittatum*). Also the related species the Citrus katydid (*Caedicia* spp.) • Leaf case moths, particularly the species often seen on citrus trees, the Leaf case moth (*Hyalarcta huebneri*) • Possums and rats, slugs or snails

Quick guide to citrus problems

Swelling or distortion of leaves or stems	• Citrus gall wasp larvae (*Bruchophagus fellis*) • Citrus leaf miner larvae (*Phyllocnistis citrella*) • Light-brown apple moth larvae (*Epiphyas postvittana*) • Very large leaves can grow when a healthy tree is regrafted, when the tree has excess nutrients, especially nitrogen, or when trees become very shaded.
Dead leaves	• Frost can cause leaves to drop as can snow or Armillaria (*Armillaria* spp.) root rot infection.
Leaf effects of nutrient deficiencies	• Boron: large dry cells in fruit flesh often with a brownish layer under the skin and thickened skins occur, leaves turn a bronze-yellow colour. • Copper: small bunched leaves dark green in colour. • Iron: leaves turn white or yellow but veins are vividly green. • Magnesium: leaves have yellow blotches between veins. • Manganese: leaves show complete yellowing between veins. • Nitrogen: leaves overall yellow and are small in size. • Phosphorus: leaves dull and leaf drop may occur. • Potassium: slight yellowing and patches of dead tissue may develop. • Sulphur: old leaves show the normal dark green but very young leaves and sometimes whole new shoots are bright yellow. • Zinc: leaves show a yellow mottling and a yellow patch in the shape of an inverted 'V' at the end of the leaf.

Fruit problems

Spots on fruit	• Anthracnose (*Colletotrichium gloeosporioides*) causes spotting or blotches on twigs, fruit and small branches. • Black spot (*Guignardia citicarpa*) shows as large black depressed spots on fruit. • Brown spot of mandarin (*Alternaria citri*) causes some shoots to die and round indented brownish spots on fruit skins to occur. • Melanose of citrus (*Diaporthe citri*) multiple tiny dot-like spots on fruit and leaves. • Septoria spot (*Septoria* spp.) shows as black dots on skin of fruit with some larger spots. Dots seem to have a scabby centre. Sooty blotch (*Gloeodes pomigena*) blotchy black greasy spots • Sooty mould (*Capnodium* spp.) shows as a sooty smudge on fruit and leaves. Can completely cover fruit.

Continued on next page

Quick guide to citrus problems

Holes in fruit	• Fruit fly larvae: Queensland fruit fly (*Bactrocera tryoni*) or the Mediterranean fruit fly (*Ceratitis capitata*) • Orange fruit borer larvae (*Isotenes miserana*) • Possums or rats
Discoloration	• Mites (see above) can cause staining of fruit • Re-greening of fruit skins from orange to green often seen with Valencia oranges can be caused by physiological or environmental conditions
Rot	• Blue mould (*Penicillium italicum*) • Brown rot (*Phytophthora citrophthora*) • Green mould (*Penicillium digitatum*) • Stem end rot (*Diaporthe citri* and *Diplodia natalensis*)
Cracks	• Citrus canker (*Xanthomonas axonopodis* pathovar *citri*) affects fruit leaves and twigs causing leaf drop, disfigured fruit, and wart type eruptions. A yellow halo can develop around the infection site on leaves. • Citrus scab (*Sphaceloma fawcetti* var. *scabiosa*) infecting leaves, twigs and fruit shows as scabby wart-like growth. Especially severe on lemon fruit skins. • Climatic changes may cause cracks and resultant rot • Copper deficiency: sudden splitting of near-mature fruit can be caused by nutritional problems (copper or calcium deficiencies), water stress, soil conditions or insect predation causing stress.
Thick skin	• Cool weather conditions • Excess nitrogen fertiliser • Thickening of skins can occur because of very vigorous growth.
Flesh problems	• Granulation (large cells filled with clear fluid) can be caused by nutritional deficiencies such as boron, or by water stress, or other environmental conditions. Fruit left on trees long after harvest period can also become granulated and the fruit skins puffy. • Spined citrus bugs (*Biprorulus bibax*) can make the flesh dry.
Distortion or scarring	• Frost can cause 'dead' or rotten fruit or fruit with shrivelled hard skins with brownish/black areas. • Katydids, light green grasshopper look-alikes, can cause scarring of fruit. • Mites (see above) can cause distortion of fruit.

Quick guide to citrus problems

Distortion or scarring *continued*	• Mutation: rarely occurring but can make unusually shaped fruit and the condition chimera, where the fruit shows two different types of skin growth on the one fruit.

Problems of bark, trunk or roots

Insects	Longicorn beetles including the Citrus longicorn (*Skeletodes tetrops*), the Citrus branch borer (*Uracanthus cryptophagus*), the Fig longicorn (*Acalolenta vastator*), the Pittosporum longicorn (*Strongylurus thoracicus*), and the Speckled longicorn (*Paradisterna plumifera*) – stem boring larvae of these and other species can bore into the wood, leaving behind webbing mixed with frass (sawdust) at the hole entrance.
Animals	Citrus nematode (*Tylenchulus semipenetrans*) can cause swellings on roots and reduce tree vigour. Rats and possums can cause problems.
Rots	Armillaria root rot (*Armillaria* spp.) shows under bark as white fan-shaped threads. Collar rot (*Phytophthora citrophthora*) rots bark at ground level. Pink disease – causes white or pink fungal growth on bark of limbs and twigs.
Soil problems	pH (a measure of acidity or alkalinity, see Glossary) can influence nutrient uptake and affect plant growth. Poor drainage will cause tree death or fibrous root death making trees show nutrient deficiency-like symptoms.
Water problems	Lack of water, too much water or irregular watering can cause problems, and severe waterlogging of root systems will kill the tree. Continuous flowering and flowering out of season can be caused by water stress.
Environmental conditions	Frost can cause the death of trees.
Physiological conditions	Alternate bearing, fruit drop, flower drop, leaf drop, lichen

Insect and Animal Pests

Ants

There are hundreds of ant species in Australia, some introduced and many unnamed. They vary in size from less than pinhead size to large bull ants 2–3cm long with a ferocious sting. Ant invasions of citrus trees are common especially if other insects such as scale or aphids are present and producing 'honeydew' on which the ants feed. Some ant species 'farm' insects such as aphids by spreading them around trees to ensure a good supply of food.

Tiny Argentine ants (*Linepithema humile*) are about 3mm long, light brown in colour with no discernable smell when crushed. They tend to move in columns along set routes. If found this ant must be reported to authorities, as must the recently introduced Fire ant (*Solenopsis invicta*) (2–6mm long, reddish brown with dark brown rear segment; very aggressive with a stinging, fiery bite). Another ant, found in NSW, is the Yellow Crazy ant (*Anoplolepis gracillipes*) that sprays formic acid to blind its prey; avoid these ants especially near eyes. If found, this ant should also be reported.

Australian green ants on mango fruit. Green ants are used as biological control agents in organic citrus groves.

Organic controls include pyrethrum-based sprays, and sticky bands around tree trunks to trap ants. We have had spectacular success controlling ants with Beat a Bug™, a product containing chilli, garlic and pyrethrum, either poured into the ants' nest or sprayed directly onto ants. Dettol™ antiseptic (diluted 10:1 in water) can be sprayed on ant pathways to dissolve track traces making the ants move on when they cannot find their scented pathway. In the Northern Territory organic citrus growers are successfully using the Australian green ant for insect control, allowing colonies to develop and stay in the tree canopy.

Aphids

There are two major species of black citrus aphids (*Toxoptera aurantii, Toxoptera citricida*) that attack young soft shoots of citrus trees, especially lemon trees. Aphids are about 1.5mm long, and both

Citrus aphids on mandarin leaves – winged and non-winged forms

winged and wingless forms are found. Aphids suck sap and cause young leaves to distort, and shoots to twist and develop abnormally. Aphids can also spread the Tristeza virus. Where aphids occur in large numbers they shed whitish coloured skins as they grow, and secrete huge quantities of honeydew (sugary excretions) on which black sooty mould fungus grows.

Soft soap or pyrethrum sprayed directly onto aphids will help control them or they can be hosed off with strong jets of water. Light oil sprays applied to the trees will smother and kill any eggs. There are many natural predators that help keep aphids under control including birds, hoverflies, lacewings, and various ladybirds. Ladybird, hoverfly and lacewing larvae are prodigious consumers of aphids so try companion planting with insect attracting and pollen supplying plants. Bird attracting flowering plants with nesting sites nearby will attract many birds.

Bronze orange bug

Mature Bronze orange bugs (*Musgraveia sulciventris*) are about 25mm long, with the typical shield shape very similar to the common green vegetable bug, but deep bronze in colour fading to bronze/black as the insect ages. The five nymph stages are fascinating involving colour variations from pale to dark green to pink or salmon colour. Nymph and adult forms of this insect are aggressive feeders, excreting an offensive smelling liquid, a good indicator of bug presence. The bugs suck sap and cause shoots to die, and leaves and flowers to drop prematurely. Their excretions have a caustic effect wherever deposited.

To control, gather and destroy insects when they cluster, pick individual insects off by hand or spray in winter with a soft soap to destroy nymphs. This is an Australian species so there are natural predators such as assassin bugs, and they are also destroyed by extremely hot or very cold weather.

Bronze Orange citrus bug – near mature adult

Citrus butterfly

The larvae (caterpillars) of two species of citrus butterfly, the large citrus butterfly and the small citrus butterfly (*Papilio aegeus* and *Papilio anactus* respectively), attack citrus leaves. The mature butterflies have a 13cm and 7.5cm wingspan respectively and are otherwise similar in appearance, although the male large citrus butterfly is less colourful than the female, but both sexes of the small citrus butterfly look similar. Both species have black, dark brown, white and grey markings with orange-red and blue markings on the hind wings. The larvae are spiky; the large citrus butterfly larvae (65mm long) are brown to olive green with brown bands and white markings, while the small citrus butterfly larvae (45mm long) are blackish with spots, and regularly placed small yellow splotches along the side. Both larvae emit a strong odour when crushed or disturbed.

Citrus butterfly (small) larva

Predator wasps and bugs help control the larvae, which rarely occur in huge numbers in home gardens. They can also be removed from trees by hand.

Citrus gall wasp

This wasp (*Bruchophagus fellis*), native to Australia, is found in the wild on some Australian citrus species. It can attack all citrus but seems more severe on cultivated lemons, grapefruit and the Rough lemon. Areas with mild winter weather sustain populations of the wasp. The black adult

Citrus gall wasp swelling cut through to show larva cells. Note that citrus gall wasp is reportable in some states.

wasp is only about 2.5mm long and rarely seen; however, the damage caused by the larvae looks hideous. The wasp lays eggs in soft stem tissue, and as the hatching larvae begin eating, they cause the tissue around them to callus or produce a more woody tissue. The resultant swelling is highly visible. In spring, when the larvae mature (about two years after the eggs are laid), exiting wasps make tiny holes all over the gall.

Galling reduces tree growth, and in severe cases when all new shoots become galled, leaf and fruit production is severely affected. Infection is usually spread from another infected tree nearby.

Wasps, ants and other predators assist with control and very hot weather at hatching time can cause many exiting wasps to die. To treat this pest, all twigs and limbs showing signs of swelling must be cut off and destroyed immediately even if this means cutting the tree right back to a stump. Check regularly for any signs of reinfection.

Citrus leaf miner

This insect (*Phyllocnistis citrella*) infested citrus trees in northern Queensland in the 1960s and since then has spread rapidly southward. It can now be found as far south as Melbourne. The larvae burrow in tunnels beneath the leaf surface of all types of citrus destroying the ability of the leaves to function properly and reducing tree growth especially in young trees. Insects are active particularly when new growth occurs on trees in spring and autumn. Tunnels in the leaves blanch silvery-white and the leaves are often severely distorted. Adults are tiny moths about 2mm long but, as they mostly emerge at night, they are rarely seen. Infestation usually comes from other trees so quarantine restrictions on tree movement are in place. Pestoil™ (the fine oil also useful in controlling various scale insects, mites and aphids) has recently been found to be an effective control if applied at times of growth flushes.

Citrus leaf miner damage, close-up view

Citrus whitefly, Australian

There are about 20 species of whitefly in Australia. The main one infecting citrus is *Orchamoplatus citri*. The tiny adult, about 2.5mm long, is a sucking insect usually found on the under-surface of leaves. Immature forms are scale-like in appearance. Whitefly multiply quickly in warm conditions to reach enormous populations on one tree. Whitefly do little permanent damage to trees, but the nymphs and adults excrete honeydew, sugary secretions that become a food source for black sooty mould (see below) and catches dust. Sooty mould reduces the efficiency of photosynthesis of leaves.

Whitefly spp. on tomato leaves, showing typical shape and colour of the species

Lacewings and ladybirds active around trees can help control whitefly. Light oil sprays such as Pestoil™ are also effective, and we have had success using Beatabug™ (see above). Pruning to open the tree canopy and allow wind ventilation can also be helpful.

Compost flies

When citrus trees are mulched heavily to conserve moisture and protect their shallow root systems, various flies (including vinegar flies *Drosophila*

Compost flies on citrus leaf

Crusader bug showing distinctive cross-marking

spp.) that inhabit mulch and compost can congregate on leaves. The flies do no harm so no control measures are necessary.

Crusader bug

This native Australian insect (*Mictis profana*) about 25mm long can be seen feeding on many plant species such as wattles, but it often invades citrus trees. It is very easily recognised: it is shield shaped with very powerful and enlarged back legs and a distinctive yellow cross on its back. Crusader bugs occur in low numbers but they can suck sap from young shoots causing the shoots to wilt and die. Removing insects from trees by hand is a simple means of control.

Earwigs

The introduced European earwig (*Forficula auricularia*) can, at times, be found in almost plague proportions depending on seasonal conditions. Earwigs feed mostly at night and shelter between fruit or on citrus leaves during the day. There is some debate about whether earwigs are pests or friendly predators, but they have been known to damage fruit, flowers and leaves of many plants by nibbling very soft tissue. One problem is that their 'frass' (excreted material) dirties fruit.

Earwig on citrus leaf

Control is not usually necessary. Moistened, loosely rolled newspaper can be used to collect earwigs when they seek shelter in the roll. The rolled paper can be collected and replaced every 2 to 3 days; this will reduce numbers. Marsupials such as antechinus eat earwigs that congregate under leaves and ground cover material.

Fruit flies

There are two species of fruit fly: the Mediterranean fruit fly (*Ceratitis capitata*) and the Australian native Queensland fruit fly (*Bactrocera tryoni* syn *Dacus tryoni*).

Fruit fly damage to 'Meyer' lemon fruit

Many years ago, Mediterranean fruit fly was commonly found in NSW and sometimes in Victoria, Tasmania and South Australia, but it is no longer an established pest except in Western Australia. It has a similar lifecycle to the Queensland fruit fly, attacking the same fruit and being controlled in a similar way. The adult fly is slightly smaller than the Queensland fly, is yellowish with black and silver markings, and the drooping wings have dark blotching and bands on them.

Queensland fruit fly is a much more serious problem, attacking soft fruit including those of many Australian plants and trees, most garden fruits and soft-skinned vegetables such as eggplant and tomato. Its presence is usually confined to Queensland and NSW coastal regions but outbreaks have occurred in Victoria. There are strict quarantine controls on the movement of fruit over State borders because of the risk of spreading this insect. When outbreaks do occur whole areas are bait-sprayed with Maldison-protein hydrolysate.

The Queensland fruit fly grows to about 7mm long and is predominantly brown in colour with bright yellow markings. It is very active, moving quickly but jerkily. Females lay eggs through a retractable 'ovipositor', a tube-like structure at the base of the abdomen. Puncture holes in fruit where the eggs have been inserted often seal or become brownish. The tiny (1mm) eggs are a distinctive banana shape hatching into creamy, 9mm long, maggot-like larvae, that can spring or jump if placed on a flat surface. As the maggots feed, the fruit flesh becomes rotten inside, although the surface of the fruit may not show any signs of damage. If green citrus fruits are 'stung', the area around the puncture hole turns yellow or orange. Fruit may also fall from the tree after being stung and fungi, such as blue mould, can invade the puncture site further accelerating rot.

Control measures include: fruit covers, Dakpot™ (bait traps used as an indicator of degree of infestation only; the traps will not kill all flies),

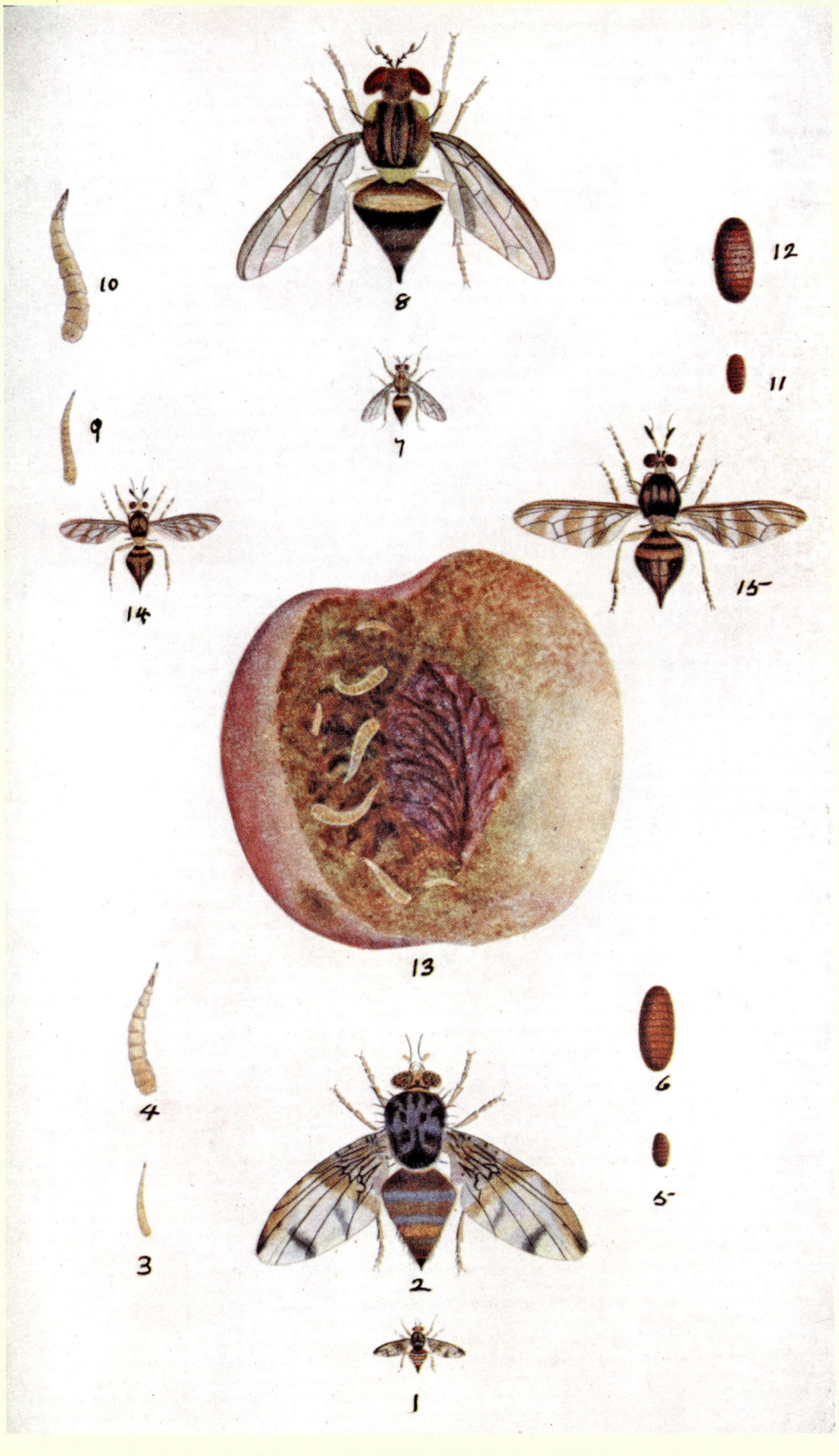

Stages of development of both the Mediterranean fruit fly and Queensland fruit fly, from pupa to adult insect with drawings of a fruit fly from the New Hebrides (actual size and magnified: 14 & 15) for comparison. The Mediterranean fruit fly larva is shown natural size and magnified (3 & 4), the pupa natural and magnified (5 & 6) and the adult fly natural and magnified (1 & 2).

The Queensland fruit fly is shown similarly: larvae in diagrams (9 & 10), the pupae (11 & 12) and the adult fly (7 & 8). The central diagram (13) shows a peach infested with Queensland fruit fly maggots.

From French, C. 1909, *A Handbook of the Destructive Insects of Victoria (with notes on the methods of prevention and extirpation)*, Osboldstone & Co., Melbourne.

insect proof netting, picking fruit before they ripen, pruning (particularly cumquat) for winter harvest when flies are not as active, using hanging sticky traps, picking up all fallen fruit and destroying immediately (see below). In some areas, release of sterile male flies has helped to keep insect numbers down.

State laws, such as in NSW, require that people with fruit trees (including citrus) must control fruit fly, which includes taking the following measures:

1. Collecting all fallen (and infected) fruit at intervals of not more than three days.
2. Destroying all infected fruit at intervals not exceeding three days by immersing in water with a film of kerosene on the surface for ten days or by burning, boiling or placing fruit in the sun in sealed plastic bags for several days to 'cook' the fruit and destroy all insect larvae.
3. Applying either poison bait spray or full cover sprays to all citrus trees. Spraying of citrus trees usually begins as the fruit start to colour.

Grasshoppers

There are many different grasshopper species in Australia but only a few may damage citrus crops. These include the Giant grasshopper (*Valanga irregularis*), the Australian Plague locust (*Chortoicetes terminifera*) and the Spur-throated locust (*Austracris guttulosa*). By far the most common grasshopper, though, is the Wingless grasshopper (*Phaulacridium vittatum*) which invades gardens when grasses and other food sources have dried off during summer. Female grasshoppers are up to 18mm long and males are about 12mm long. Rarely are fully winged specimens found. Grasshoppers are brown-tan or greyish, with

Wingless grasshopper

some being black in large populations. The insides of the large back legs are usually bright red.

Grasshoppers nibble leaves and, in large numbers, can strip everything bare. They hatch in spring, go through several moults (or instars), lay eggs during summer but die by early autumn relying on the next generation to continue the species.

Chemical control often seems the only option, but organic growers and permaculture exponents employ chickens in the orchard to control numbers. Attracting insect-eating birds is also an option, although birds can also damage fruit. Insect proof netting is another option. Neem, a plant oil, may be available within a few years for grasshopper control.

Katydids

Katydids belong to the long–horned grasshopper family *Tettigoniidae*. There are two species, the Inland katydid (*Caedicia simplex*), and the Citrus katydid (*Caedicia strenua*). There are slight differences between the species in abdomen colour, overall size, green colour and ovipositor appearance. They are about 45mm long, green, with upward folded wings forming a triangle shape over the abdomen, and long antennae. Females of both species have a short upward curving ovipositor (egg laying apparatus). Katydids can change

Katydid skin damage to 'Valencia' orange fruit

colour from green to purple-pink seemingly as a result of their food source or the colour of the foliage in which they are sheltering.

Katydids cause little damage to citrus foliage but can rasp away at the skin of very young citrus fruit creating a sunken, scarred area, at first whitish but turning speckled as fruit matures. This scarring is very visible and in commercial orchards can make the fruit unsuitable for sale. Katydids generally invade one tree at a time but not in very large numbers, making hand removal an effective method of control.

Leaf case moth

There are many native Australian case moth species found in home gardens, but one in particular, *Hyalarcta huebneri,* has been identified as able to

Leaf case moth species larvae in casing on plane tree leaf

cause serious damage to citrus tree foliage. Its larvae spend their whole life inside a silken webbed pocket that is extended as the larva grows, eventually extending to as much as 65mm. The outside of the case is covered with pieces of citrus leaf. The wingless female pupates inside the casing and a winged male will mate with her. Eggs are laid in the case and newly hatched larvae leave through the bottom of the case. Larvae eat holes in leaves, often only eating through one layer, leaving transparent patches not unlike snail damage. Often this insect is only present on one tree at a time, so it can be controlled by hand removal.

Light-brown apple moth

This moth (*Epiphyas postvittana*) is usually associated with apple trees, but it has a huge host range including citrus trees. Moths have a wingspan of almost 2cm and are brown-grey in colour. Larvae are bright light green and very active when disturbed. Larvae spin webbing, often between folded leaves for shelter. They eat leaves, chew holes and can cause damage to young fruit, scarring the stem end when the skin is soft. The biological agent, *Bacillus thurengiensis*, commonly available as Dipel™, is an adequate control. Twist-on ties for control are available as well. Weed control is also important as weeds can play host to the moth.

Light brown apple moth (LBAM) damage to citrus leaves

Longicorn beetles

There are many longicorn beetles native to Australia, some of which are still to be identified. Several have been found in citrus trees, including the Citrus longicorn (*Skeletodes tetrops*), the Citrus branch borer (*Uracanthus cryptophagus*), the Fig longicorn (*Acalolepta vastator*), the Pittosporum longicorn (*Strongylurus thoracicus*) and the Speckled longicorn (*Paradisiema plumifera*).

Longicorn beetle larva (typical form with squarish head)

These mostly nocturnal beetles usually lay eggs on dead or dying bark of sickly trees or around wound sites. All longicorn beetles have distinctive curved antennae often longer than the body that is usually thin and narrow. Many of these beetles make a buzzing noise if handled. The roughly triangular shaped segmented larvae are also unusual in that they appear square headed when viewed from the front. Larvae are found in tunnels either just under the bark of trees or in holes in central wood of branches where they pupate after feeding. Adult insects can be seen emerging from trunk or limbs. Attack by these beetles is usually sporadic and only affects already weakened trees. Wasp parasites infecting larvae help with control as does maintaining tree health and ensuring that wounds are treated (see Chapter 3 and Glossary).

Mealybugs

Several mealybug species are found on citrus trees including Long-tailed mealybugs (*Pseudococcus longispinus*), Citrus mealybug (*Planococcus citri*) and Citrophilous mealybug (*Pseudococcus calceolariae*). Mealybugs are soft-bodied sapsuckers, usually powdery white coloured. They are small (only about 6mm), oval and segmented like slaters. Mealybugs have short filaments all around and some species have longer tail filaments. They exude a sticky substance especially when disturbed, and this substance spoils fruit and provides food for sooty mould. Mealybugs can be attacked by parasites and predators such as lacewings and ladybirds so both can be encouraged. Pruning trees to allow good wind movement and light through the foliage will discourage mealybugs. Pestoil™ will have some effect on populations. Small numbers can be removed from branches and foliage with a fine stiff brush or by pruning off and burning infected parts.

Mealybug – close-up showing powdery coating and filaments of insect

Mites

There are a number of mite species that affect citrus.

Citrus bud mite (*Eriophyes sheldoni*): The Citrus bud mite is an eriophyid mite that has a spectacular effect upon some citrus fruits particularly lemons. The mite will attack all soft tissue and cause some distortion in shoots, resulting in fan-shaped shoots and rosetting of terminal growth (see Glossary) and leaves, but its effect on fruit is the most significant: lemons can

Rust mite on sunny side of sweet orange fruit (brown citrus rust mite)

become completely distorted with lumpy, 'finger-like' growth occurring all over the fruits.

The tiny mites (about 0.17mm long) can just be seen with the aid of a ×-10 power hand lens. The sausage-shaped adults are cream or pinkish coloured with two sets of legs. The mites are active most of the year and can be found on new shoots and buds and flowers. Lime sulphur or dispersible sulphur sprays or low dose oil sprays will give some control.

Citrus rust mite (Maori mite) (*Phyllocoptruta oleivora*) is a pest of citrus trees only. Citrus rust mites are tiny (0.15mm long), yellowish, slender with two pairs of legs and a pair of 'false' legs at its base. In comparison, Brown citrus rust mite (*Tegolophus australis*) is about 0.18mm long, almost wedge-shaped and brown or light brown.

Staining on lemon fruit caused by citrus rust mite

Rust mites can be seen on leaves and shoots and feed on green undeveloped fruit; damaged skin shows as mottled or streaky brown (Citrus rust mite) or a shiny polished brown (Brown citrus rust mite) colour when the fruit ripens. Citrus rust mites damage skin on the sheltered side of the fruit and fruit that is under leaf cover whereas Brown citrus rust mite damage is to the exposed section of fruit skin. Mites are partially controlled by natural predators such as ladybirds. Sprays of low dose oil sprays or lime sulphur or wettable sulphur have some effect on populations of these mites. Chemicals for mite control are also available.

Broad mites (*Polyphagotarsonemus latus*) have been known to cause damage to vegetables and many ornamental plants. They can invade citrus trees, particularly targeting young fruit or newly developed foliage. Most citrus do not suffer severely but lemon trees that set fruit in the late summer period are susceptible. Mites damage the fruit until they reach about 3cm in size and can turn lemon skins silvery as a result of the injury. Sometimes spotty rough areas appear on the skin with a bronze or grey colouring. Control is as for other mites.

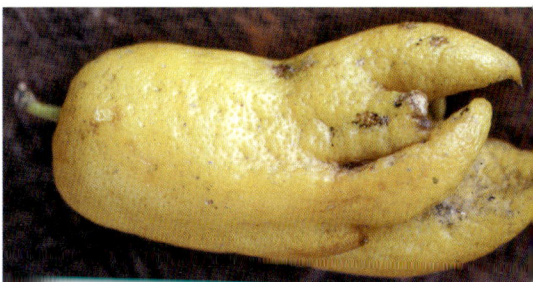

Citrus bud mite of lemon fruit, causing a 'claw' shaped fruit to develop

Citrus red mites (*Panonychus citri*) are about 0.5mm long, oval with four pairs of legs and are a dark red or purplish red colour with a bristly appearance. Eggs are also bright red in colour. These mites attack young leaves and citrus fruit, and can be seen sometimes clustered on trunk bark. Leaves, fed on by mites, have a washed-out appearance and fruit can be blemished. In some instances, leaf fall may occur. Predatory mites and ladybirds can help control this mite and other organic (see above) and chemical control measures are also an option.

Orange fruit borer

The small adult moth (*Isotenes miserana*) is grey coloured with mottled brown markings with a wingspan of about 15mm, and the larvae of this small moth cause injury to ripening citrus fruit although rarely to citrus foliage. Larvae turn from green to a cream colour with a brown head and are very active if disturbed. There is usually only one grub (larva) found in a single fruit: they bore through the outer skin to the pithy section on which they feed. These insects can be controlled chemically, but the best way to reduce numbers is to destroy any fruit with telltale holes. Thinning of fruit, so that they don't cluster tightly together, can also help with control by reducing the number of sites where larvae can hide.

Imperial mandarin, showing orange borer damage also webbing and skin damage by Light brown apple moth (LBAM)

Passion vine leaf hopper

This Australian sucking insect (*Scolypopa australis*), about 8mm long, looks like a moth with brown, mottled, see-through wings. It can invade a large range of garden plants including passionfruit, ornamentals and grape vines. Young nymphs are small with upright 'peacock' fluffy white tails. Both adults and nymphs can hop and jump.

Young Passion Vine Leaf Hopper (PVLH) on Dietes flower stem (to show young hopper stage)

Adults at rest have their wings spread. They exude colourless honeydew that will attract sooty mould. Populations may be severe with hoppers clustered on every section of new growth. Although they do not generally cause much damage, new shoots that have been attacked will wilt. Predators and birds, such as sparrows, will feed on them. Attracting birds to the garden will help as will spraying with organic, pyrethrum-based sprays.

PEST AND DESEASE CONTROL OF CITRUS

Citrus fruit ('Meyer' lemon), still hanging in tree after skin was removed by possums

Possums

The brush-tailed possum and the ring-tailed possum can both be found in gardens and orchards. Brush-tailed possums will eat skin from citrus fruit and are especially partial to lemons; they can eat all the skin without making the fruit fall leaving the skinless fruit hanging by its stalk. Possums, although often a pest, are protected so check with relevant State wildlife experts for control measures.

Rats

Rattus rattus can damage citrus fruit in a similar way to possums and they will also attack stored citrus fruit if they can get access to them. Rats can be excluded from food sources, trapped, kept under control with dogs or cats, or controlled by licensed pest controllers. Rats are attracted to food scraps and poorly managed compost heaps, so having clean and tidy well-managed orchards is essential.

Navel orange fruit damaged by rats

Scale insects of citrus trees

There are a number of species of scale that affect citrus including Black (or Brown olive) scale (*Saissetia oleae*), Californian red scale (*Aonidiella aurantii*), Chinese wax scale or Hard wax scale (*Ceroplastes sinensis*), Citricola scale (*Coccus pseudomagnoliarum*), Circular black scale or Florida red scale (*Chrysomphalus aonidum*), Cottony cushion scale (*Icerya purchasi*), Mussel or Purple scale (*Lepidosaphes beckii*), Pink wax scale (*Ceroplastes rubens*), Soft brown scale (*Coccus hesperidum*), White louse scale or Citrus snow scale (*Unaspis citri*), White wax scale (*Ceroplastes destructor*), and Yellow scale (*Aonidiella citrina*).

Scales are tiny sucking insects belonging to two different groups: armoured scales and soft scales. Soft scales are generally larger but both types have tough resilient coverings, soft as with the native Australian cottony cushion scale, or hard scaly covers typical of the Californian red scale.

Scale insects are by far the most important pests of commercially grown citrus trees and are often found in home gardens. Although scale insects do little harm to fruit and foliage, they create blemishes and their honeydew (sugary secretions) provides sites for sooty mould to develop. Citrus fruit for export must be free of scale insects but home gardeners should not worry about mild infestations except where control is

Chinese Wax scale on citrus tree branch

Citronelle fruit with Californian red scales on skin

Cottony Cushion scale on citrus tree lateral

Citrus fruit showing skin mutation (ridged) and Californian Red scale.

mandatory. Tiny six-legged scale nymphs (crawlers) usually settle in one place soon after hatching, develop their outer covering and seldom move as adults.

Immature scale may have several moults. Armoured scales lose their legs at crawler stage during the first moult, while soft scales may retain their legs to adulthood in some species. Fully developed females usually remain static on plants, to be fertilised by flying males with a very limited life span. Some scales seem to be parthenogenetic, producing young without fertilisation occurring. Females lay eggs under their scale covering and hatch them inside their bodies.

Controls vary with the species. It is the crawler stage that spreads infection as crawlers can be blown by wind, travel on the legs of birds, be farmed by ants, or distributed in a range of other ways. Organically grown citrus trees have many insects and animals on them and some are predators of citrus scale, including wasps (parasitic and predator), hover-flies, ladybirds, lacewings, earwigs, preying mantis, assassin bugs, and some native caterpillars. Birds feed upon scale insects. Parasitic wasps are keeping many of the scale insects under control on Australian citrus orchards.

Integrated measures to control pests and diseases are applied with great success; generally in combination with biological controls, less chemicals need to be used. Specific control recommendations for each scale species are provided by State Departments of Agriculture, but as a general rule, light oil sprays (e.g. Pestoil™ or Eco oil™ used when crawlers are active will reduce scale insect populations significantly. In China, where labour is very cheap, the only pest control measures taken are to meticulously spray or apply a 1% solution of oil to the whole tree. Other organic control methods include predatory wasps (available via mail order – see Resources section), soap sprays, pruning out and destroying infected sections, preventing ants climbing trees with 'sticky' bands, and companion planting to encourage predator populations.

Slugs and snails

These are often found in and around citrus trees. Slugs are a relatively minor pest but can damage citrus fruit especially when at or near ground level, and they may chew soft new leaves of seedlings or grafted plants. The giant leopard slug (*Limax maximus*) is a predator of other slugs, rather than a pest, and should be left alone.

Common garden snail (*Helix aspersa*)

Orange fruit borer damage and slug on mandarin fruit

The common garden snail (*Helix aspersa*), on the other hand, can become a serious pest. Snails are very active during warm wet periods and can attack young shoots, leaves and fruits of citrus trees. Typically, leaves have holes with some 'see-through' sections, and fruit may have holes in the skin which then develop rot. At rest, snails tend to cluster on tree trunks and are easily collected. Some birds and small mammals eat slugs and snails, and give partial control, but hand collection, the use of beer traps and the organically acceptable Multiguard™ pellets will control populations.

Spined citrus bug

The Australian native Spined citrus bug (*Biprorulus bibax*) is about 2cm long and commonly found on citrus trees in NSW and Queensland. Otherwise resembling the common green vegetable bug, this bug is unusual because it has two short, sharp horns on its head. Nymphs are rounded and can be yellow-green, black and orange, or green mottled with black gradually becoming fully green colour at the adult stage. Citrus bugs feed on fruit, causing slight skin damage and the cells inside the fruit to dry out. Small fruits may yellow and fall from trees. Gumming and browning may occur inside damaged fruit, especially where the bugs' sucking organs have penetrated. Seeds sometimes shrivel. The bugs often cluster on trees and do not fly during early winter so they are easy to collect by hand and destroy.

Spined citrus bug adult and nymphs, showing damage to leaves caused by these sucking insects

Stem borers

There are various species of stem borers, many of them native to Australia. Larvae burrow into, and often around, limbs removing bark layers. Some borers attack citrus tree branches causing girdling (removing bark right around the circumference), effectively ring barking the branch resulting in its death. Damage is easily seen and often associated with a mass of webbing and frass (sawdust) around the wound area. Cutting off the damaged branch, and destroying or burning it to kill the larvae, prevents further damage to the tree. Poking a wire into holes to skewer the larvae is another option.

Stem borer damage to citrus branch

Thrips

There are a number of thrip species including Western Flower thrip (*Frankliniella occidentalis*), Greenhouse thrip (*Heliothrips haemorrhoidalis*), Citrus rust thrip, also called Orchid thrip (*Chaetanaphothrips orchidii*), Plague thrip (*Thrips imaginis*).

Thrips are tiny (about 1–1.5mm long) and to the naked eye look like flecks of material. They are pointed at both ends, long, thin and narrow. Colour varies, but newly hatched nymphs are usually white or yellow, adults are black or yellow and may have markings. Thrips are found on many garden plants especially flowers, and often in large numbers. They can be blown from one place to another by wind, transported or carried by insects or birds.

The rasping mouthpart of thrips can cause physical damage when they feed. Citrus fruit shows discoloured patches that turn soft when the fruit is stored. Petals and the soft material in flower parts may callus or blister, or show some browning that dulls with age. There are some wasps and other predators of thrips but these are not very effective when thrips are in large numbers. Soapy water or oil or pyrethrum sprays are good control measures, as are white or pink sticky hanging traps. Very hot or very wet conditions do not favour thrips, so severe infestations will not occur every year.

Thrip damage to leaves of citrus tree

Weevils

Those found infecting citrus include the Citrus fruit weevil (*Neomerimnetes sobrinus*) Apple weevil (*Otiorhynchus cribricollis*), Citrus leaf-eating weevil (*Eutinophaea bicristata*), Spinelegged citrus weevil or Dicky rice weevil (*Maleuterpes spinipes*), Elephant weevil (*Orthorhinus cylindrirostris*), Fruit tree root weevil (*Leptopius squalidus*), Fuller's rose weevil (*Asynonychus cervinus*), and Whitestriped weevil (*Perperus lateralis*). Of these, the most damaging are the Elephant weevil and the Fruit tree root weevil.

Weevils generally lay eggs in tree bark or in the soil, and it is the larvae stage that causes most damage. Most weevils have a protrusion or beak from which antennae grow. Some species have retractable beaks. These beetle-like insects have a very hard shell.

Australian native Elephant weevils (*Orthorhinus cylindrirostris*) will attack most orchard fruit trees and will also damage grape vines. Fully mature insects are 1–2cm long, grey/black in colour with a scaly covering and the typical protruding antennae. Creamy larvae, up to 2cm long and legless, tunnel in roots causing injury, and sometimes they ringbark roots, allowing diseases to take hold. The insect pupates and then emerges as an adult, boring an exit hole in the trunk or limb. Adult weevils remove bark from twigs and branches.

Control of weevils is difficult, but keeping trees healthy and well-watered to prevent stress will help prevent weevils taking hold. Sticky glue bands around tree trunks will prevent weevils climbing the tree, and damp, scrunched up newspaper around the base of the tree is good for collecting and destroying weevils.

Fruit tree root weevil (*Leptopius squalidus*) can cause die-back in citrus trees. They are about 2cm long, grey, with a shortened protrusion or beak and antennae, and do not fly. Larvae – about 2cm long, legless and creamy coloured – eat deep channels in root surfaces. Sticky trunk wraps, and keeping weeds low prevents them climbing trees (via tall weeds). Damp newspaper, as suggested above, helps collect weevils.

Root borer (Fruit tree root weevil) adult on peach tree branch

Diseases of citrus trees

Most diseases of citrus are fungal in origin although some are bacterial. Diseases of citrus have been documented in Australia from the time of the earliest agricultural advisory services. In 1899, the Victorian Department of Agriculture pathologist listed the major diseases of the time including 'False Melanose, Anthracnose (Black Spot), Sooty Mould of Orange and Lemon, Black Scurf of Citrus Fruit, Scabbing of Fruit and Leaves, Wither Tip (Brownish-black Scab), Bark Blotch of Lemon, Collar Rot, and Root Rot of Lemon'.[26]

Accompanying the descriptions of these fungi are descriptions of other fungi, both parasitic and saprophytic, that had been found on citrus fruit (21 species), citrus leaves (37 species), citrus stems and branches (20 species), fungi on roots (one species), and fungi on scale insects of citrus (three species). Only a few of these fungi pose any real problem for home garden citrus growers today.

Anthracnose

Anthracnose

Although not a major disease problem this fungus (*Colletotrichium gloeosporoides*) can cause spotting or blotch marks to appear on fruit, and small branches may also be attacked causing die-back.

PEST AND DESEASE CONTROL OF CITRUS

Ladybird resident on Imperial mandarin leaf

Friendly insects: preying mantis on tansy flowers (companion plant)

ORGANIC MANAGEMENT

Good organic management techniques, including weed control, use of organic based manures, adequate drainage, encouraging or use of biological control methods, pruning out infected material, air flow through tree foliage, adequate watering and appropriate organic sprays (including copper and oil sprays) are enough to keep most trees clear of infection.

Organic control measures: cylindrical 'sticky traps'

Control of this disease is by pruning and destroying all weak and diseased branch sections and fruit. Regular copper sprays also help.

Armillaria root rot

Armillaria (*Armillaria* spp.) is often called 'honey fungus' because of its honey-coloured 'mushrooms', or 'shoestring fungus' because of the black/brown growths that form on the outer surface of the bark at the infected site. These growths can spread the disease from plant to plant or within the soil. The fungus also forms flattened, fan-like creamy white mycelium (fungus roots) between the bark and inner wood. Spores produced by the 'mushrooms' can spread the disease and old stumps are excellent sites for fungi to infect. Infected roots or plant tissue taken from one site to another will also spread the disease.

The fruiting bodies of *Armillaria luteobubalina* (one of the Armillaria species)

Armillaria infects many native plant species and ornamental plants. On citrus, the infection causes yellowing leaves, leaf fall and gradual loss of tree vigour.

Control is difficult although burning of tree stump branches and roots on site can help. Soil sterilisation, removal and replacement of soil and chemical treatments have all been tried. On new orchard sites it is important to clear all the land of old stumps and roots and burn them. Building biological activity in the soil around infection sites before replanting can also help.

Black spot

Black spot (*Guignardia citicarpa*) can be a serious fungal disease in tropical and semi-tropical areas, mainly infecting fruit but leaves may show sunken spots. The fruit symptoms usually start to show during warm weather conditions and vary having four distinct forms:
- Speckled blotch develops on green fruit in late summer causing brown-black speckling on citrus fruit skin.
- Freckle spot causing depressed spots to 1mm wide that are a deep orange to brick red in colour. The enlarged form creates hard spot (see below).

Two lemon fruits, one showing frost-damage, the other Black spot disease

- Hard spot produces roughly circular depressed spots to 3mm wide with black margins and grey-white centres. On mature fruit there may be a green margin around the spots.
- Virulent spot: at this stage the spots grow rapidly and can coalesce, creating very large depressed spots with dark fruiting bodies giving the spots a black appearance.

Spores ejected from the fruiting bodies infect fallen leaves and remain a source for further infection. Black spot can be controlled using copper and oil sprays combined.

Lemon showing Black Spot disease

Brown rot

This disease organism (*Phytophthora citrophthora*) is responsible for both citrus tree collar rot and root rot. The fungus has spores that can move in water so it can be spread by rain or splashed water, particularly water that has touched infected soil. Brown rot particularly affects lemons, and shows as a spreading brown discoloration that can become less brown as the disease spreads. Areas injured by brown rot are likely to be invaded by blue or green mould. The rot gives off a very distinctive pungent odour. Copper sprays are usually effective, but pruning the skirt of the foliage area (see Chapter 5 on pruning) and opening the canopy to allow air flow will also help.

Lisbon lemon fruit showing citrus mite damage and brown rot

Mandarin fruit infected with brown spot of citrus

Brown spot of mandarin

This is a fungus disease (*Alternaria citri*) particularly common on mandarins, although other citrus such as tangelos and grapefruit may be attacked. The disease thrives in cool damp conditions especially where foliage dries slowly. Small, light to dark brown/black indented, almost circular markings on stems of shoots or on fruit, grow to about 3cm in diameter. Leaves show a 'water-soaked' appearance at infection sites, and may curl or fall. Citrus shoots can be killed.

All infected or dead material should be removed, then burnt. Keeping the area under trees clear of debris is also advisable.

Citrus canker

In June 2004, an outbreak of this bacterial disease (Citrus canker *Xanthomonas axonopodis* pathovar [pv.] *citri*; Canker A) occurred in a Queensland citrus grove.

If it were to become established, it could have severe consequences for the citrus industry. Several variants of this disease have been found including Canker B *Xanthomonas axonopodis* pv. *aurantifolii*. Some disease variants seem to infect only a few species of citrus, but canker A has a huge host range that includes most of the common citrus grown by home gardeners.

The bacteria can infect leaves, stems and fruit and is spread by rain runoff, wind, overhead irrigation water and dew. Any injury to leaves, stems or fruit caused by pruning, wind damage or frost, will provide infection sites for this disease as will tissue injury caused by the citrus leaf miner and other injurious insects. Workers in an infected orchard or nursery can accidentally spread the bacteria by the movement of plant material or infected fruit and by scions and budwood that are infected. Machinery such as pickers, pruning saws and fruit bins can spread this disease as can small tools such as those used for budding and grafting. Stem lesions have long-term viability as will any infected material that is stored in dry conditions with no contact with soil particles.

The disease disfigures fruit showing as cankerous eruptions of tissue, and crater-like scabby lesions that may coalesce. Leaves have soft white, outward growing, wart-like growths that turn brownish. Infection produces a watery glazed area around the lumps that can be brown, green or yellow coloured with a yellowish halo of tissue around the infection site. The disease causes leaf drop and defoliation, reducing crop production severely (see Glossary).

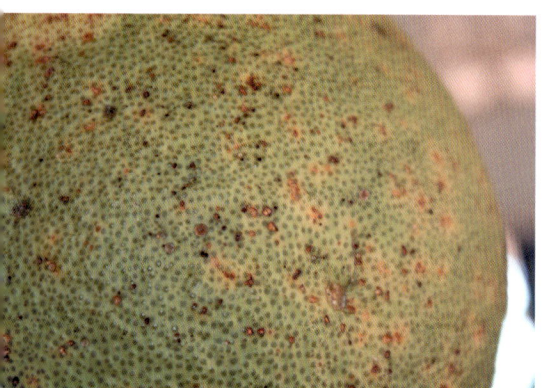

Left: Citrus canker, Pomelo fruit – Mekong Delta, Vietnam
Right: Citrus canker, Pomelo leaf – Mekong Delta, Vietnam
courtesy Bob Nissen, Department of Primary Industries and Fisheries, Nambour, Queensland.

Controlling this disease is very difficult. Bordeaux (a mixture of copper sulphate and lime) has been tried but the only effective option is to remove and burn trees and follow quarantine regulations. If the disease becomes established, resistant cultivars will have to be developed to replace some of the existing plantings. If the disease were to invade cooler areas in Australia, the spread would be rapid as it prefers damp, cool conditions.

Since the 2004 outbreak, tree removal and eradication procedures for all citrus species including those in home gardens have been put in place, and strict quarantine procedures have been implemented, including controls on movement of fruit or citrus plant material. If found, this disease must be notified to the relevant State departments of agriculture.

Citrus scab (Australian)

This fungal disease (*Sphaceloma fawcetti* var. *scabiosa*) infects citrus leaves, twiggy growth and fruit. It is most noticeable on lemons and can infect all lemon cultivars, as well as sour oranges, limes, some mandarins and tangelos. It does little damage to sweet orange cultivars.

Twiggy branches of severely infected trees become unproductive and inefficient. Bark is injured and mottling, spotting and yellowing of adjacent leaves prevents photosynthesis (see Glossary). Fruit has very visible corky, wart-like lumps and small scabby areas on the skin. Areas under and around infection sites on green fruit turn yellow. The warty

Fruit of lemon showing citrus scab disease

growths produce pink, grey or brownish scabs, and these produce spores then carried elsewhere by wind-driven rain, insects or birds.

All diseased material must be pruned from infected trees and burnt. Copper-based sprays such as Bordeaux can also help.

Collar rot (see also Brown rot)

The fungus (*Phytophthora citrophthora*) that causes this disease is soil-borne and can attack citrus tree trunks depending on conditions. Trifoliata and citrange rootstocks are often used for citrus because they are resistant to collar rot, but if mulch or weeds cover and dampen the scion above the graft then collar rot can still infect trees. To prevent collar rot, keep the under-tree area free of weeds, prune low branches, avoid wetting the trunk and ensure good drainage.

Citrus tree showing bark removal to repair collar rot disease

Tree trunk painted with Bordeaux paste to fix collar rot of citrus

The first signs are a gradual yellowing of leaves, which may eventually fall if the disease is not treated quickly. The disease attacks the bark at or near ground level causing it to rot; it can spread around the trunk ring-barking the tree.

Using a sharp knife, all infected bark must be scraped from the trunk back to the hard inner wood even if the infected area is below soil level. Bark must be scraped away until healthy tissue is found. Edges of the bark must be carefully chamfered (see Glossary) to aid healing. Infected areas below soil level, once treated should be left open to the air; no soil or mulch should be placed over the area.

All infected bark must be burnt, and exposed areas should be covered with a thick Bordeaux paste (see Glossary). Small dead limbs or twigs should be pruned from the tree and the tree should be given a complete fertiliser or animal manure (such as fowl manure) making sure none touches treated areas. Bark will regrow over scraped areas and if they are kept clear of soil, weeds or mulch, reinfection should not occur.

Greasy spot

This condition, found sometimes on the back of citrus leaves, is thought to be caused by a fungus related to that which causes Melanose rot of citrus fruit (*Diaporthe* spp.). Slightly raised lumpy spots of what looks like grease appear on the backs of leaves, which may show some curling.

The condition may be associated with disease, insect or mite damage or it could be an environmental effect, with weather or water playing a role in its appearance. The greasy spot does not seem overly contagious and only occurs on a few leaves at a time, but can cause some leaves to drop from the tree.

Greasy spot on mandarin leaves

Melanose of citrus

This fungus disease (*Diaporthe citri*) shows various symptoms, the most noticeable being pinprick dots covering leaves or sections of fruit. When there are many dots on leaves there can also be an associated yellowing. Dots are usually a dark red/brown in colour so stand out against leaf background colour. Spores of the fungus are spread by moisture such as rivulets of water associated with rain, wet fog or heavy dew. Infection can spread in finger-like patterns following water movement. With severe infection, dots on fruit may coalesce into patches, which may then crack.

This disease is often seen on older trees with large foliage areas where there are many sites (dead and fallen twigs and dead branches) for the fungus to live. All citrus are susceptible, but some sweet orange cultivars, specific mandarins such as 'Emperor', and grapefruit and lemons particularly so.

Pruning out all infected material, all dead branches and twigs and keeping debris away from trees are good preventive measures, and Bordeaux sprays can help with control.

Lemon fruit with skin infected with Melanose disease

Penicillium

The two fungi, Blue mould (*Penicillium italicum*) and Green mould (*Penicillium digitatum*) are often seen in association with one another. They invade injured tissue, even where there are minute injuries that cannot be seen with the naked eye such as skin cell breakage. The first sign is a water-soaked area with the mycelium (fungus roots) growing in a spreading circular pattern. Behind the mycelium, spores start forming; with blue mould these are powdery-blue while green mould produces distinctive greenish-coloured spores.

Rough lemon infected with Blue penicillium mould

Navel orange infected with Green penicillium mould

Blue mould is a contact mould: it can spread from infection sites by touching other uninjured fruit. If one diseased fruit is present, the disease is capable of spreading quickly especially when fruit is stored in containers. Green mould is similar in habit but seems to grow more quickly. It has a wider mycelium band at the edge of the water-soaked area of rot; this mycelium band is pasty whereas that of blue mould is powdery.

Being careful will help with prevention and control of these two moulds. Avoid damaging fruit when picking, handling, and storing, as even a fingernail puncture can start a rot. Wearing cotton gloves when picking citrus fruit, cutting fruit stems with sharp clippers to avoid pulling out stem pieces, are good preventive measures, as is avoiding picking fruit when wet. Using bags made of soft material for picking, and avoiding bruising of fruit will also help prevent rot setting in as will placing picked fruit in shady areas to avoid sunburn. Fruit should be kept cool at, during and after harvest. It is very important to collect and dispose of fallen or diseased fruit before picking begins, as these are a source of fungal spores that can cause infection.

Pink disease

This fungus (*Corticium salmonicolor*) can be quite serious during wet weather conditions, particularly in tropical and semi-tropical areas. It causes wilting of twigs or small branches and shows as a white or pink growth on the bark that can girdle the twig causing it to die. After infection the bark splits and the disease can gradually spread throughout the tree, causing grey-coloured dead twigs to appear. It is important to control the spread of the fungus by pruning out all infected material and destroying it.

Septoria spot

Where trees are wet during watering, this fungal disease (*Septoria* spp.) may become established. In citrus groves where overhead sprinklers were used it was prevalent; however, it has beome less common because of trickle irrigation and under tree sprinklers. Disease spores germinate in autumn and then become semi-dormant until wet winter conditions prevail. It shows as black dots on fruit skin, with some dots larger than others with a grey, scabby growth at the centre.

To reduce the chances of infection, prune out all dead or unthrifty wood or foliage and spray with a copper spray such as Bordeaux (see Glossary) during the March period.

Septoria spot

Sooty mould

Sooty moulds (*Capnodium* spp.) develop on leaves, fruit and stems of citrus, making the area of infection look like it is covered in black soot. Sooty mould feeds on the 'honeydew' exudations of insects such as aphids and scale. Similarly, sooty blotch (*Gloeodes pomigena*) lives on exudations causing blotchy spots to develop on trees. These sooty moulds are, however, superficial diseases not harmful except by excluding light from leaves and making fruit uninviting.

Grapefruit and foliage showing sooty mould and sooty blotch fungal growths

The best control option is to control the insects such as aphids and scale, then sooty mould will disappear. Low dose oil or copper sprays such as Bordeaux can be used but this is not generally necessary.

Stem end rot

Stem end rot on Tahitian lime fruit

These fungi (*Diaporthe citri* and *Diplodia natalensis*) are spore-producing disease organisms. Spores can settle on fruit then germinate during wet weather, remaining dormant until fruit starts ripening. Initially, the green 'button' at the base of the fruit stem is infected, but then infection, visible as a characteristic light tan colour, spreads down the fruit.

These fungi also cause die back of twigs and small branches. Making sure the tree is healthy with all dead or dying branches or twigs removed and disposed of, will prevent the disease taking hold.

Non-pathological disorders of citrus

These are problems caused by environmental conditions such as changes in weather, or physiological behaviour of plants.

Alternate bearing

Some citrus such as mandarins, cumquats and Valencia oranges tend to bear well one year but not the next. This physiological condition may be influenced by factors such as moisture supply and nutrition. If sweet orange trees, for example, are not fertilised or watered adequately during the year, shoot growth is reduced and because sweet oranges bear fruit on new growth, the following crop is reduced. To reinstate regular fruiting, prune during the 'on' year when a large crop is predicted.

Flower drop

Most citrus trees have an overabundance of flowers and if every flower produced fruit, trees would be overladen with very small fruit. It is interesting to note that only 1–4% of all the flowers on a tree will set fruit that reaches full maturity. Trees respond by dropping some flowers, and even some developing fruit.

Frost

Frost may cause severe damage, even to mature citrus trees. In 1895, for example, freezing weather in Florida in the USA virtually wiped out that citrus industry overnight. Fortunately, in Australia, it is rare to experience serious frost but it is worth being prepared nevertheless. Smudge or fuel pots, interspersed through a citrus grove, have been used to heat the air and prevent frost damage. Large, propeller-driven wind machines are used to circulate warm air to prevent frost effects. Activating sprinklers results in water on fruit and foliage turning to icicles as the temperature reaches freezing point. Ice melting in the morning sun, gives off heat and prevents frost damage to plant tissue.

If struck by frost, all leaves will curl and can gradually fall off the tree. The only way to save the tree is to severely prune it, after the danger of frost has passed, and paint the bark area of the surviving stump with water-based white paint for protection from sunburn.

Fruit drop

Many citrus species, for example grapefruit (*Citrus* × *aurantium* syn *C. paradisi*), have a habit of dropping fruit from the tree before it is ripe enough to pick. This is mainly due to physiological and weather conditions influencing abscission of fruit from the tree. Home gardeners can prevent fruit drop by adopting good management techniques so as not to stress the trees. The application of liquid seaweed products and fish food emulsion products as foliar sprays to the trees may help reduce fruit drop.

Fruit drop also occurs during and just after the period of fruit set. Immediately after pollination and when fruit is tiny, small developing fruit may drop from the tree. This is mainly due to an overload of fruit, poor nutrition, water stress and other environmental factors, and part of a natural shedding. Sometimes fruit will also drop from trees when fully ripe or near maturity. Generally, it is nothing to worry about, as plenty of fruit will be left on the tree.

Granulated flesh

Individual cells within each slice of citrus fruit can become granulated and dry without much juice. This condition can be caused by several factors including boron deficiency (see nutrition section, below) and by sucking insects. Fruit left on trees too long will also develop to this stage, as will trees suffering the effects of drought, water stress or poor nourishment. Keeping trees well watered and nourished should prevent symptoms occurring. See photo, page 92.

Greening of fruit

Fruit skin sometimes regreens after ripening, particularly some Valencia orange cultivars. Commercially, fruit is placed in a sealed room and given ethylene gas in order to recolour the fruit skin before sale. If home gardeners want to recolour fruit, all you have to do is place the citrus fruit in a sealed plastic bag containing some ripening apples or bananas and the citrus fruit skin will again colour because the ripening fruit gives off natural ethylene gas. Some citrus cultivars including some limes do not colour completely when ripe, especially when grown in tropical and semi-tropical climates.

Leaf blotch

Leaves are sometimes subjected to sudden changes of environment or temperature, and the oils cells of the leaf can rupture causing orange-coloured spots on the underside of the leaf. Leaf curl is often associated with this condition and the leaves usually become yellow and fall from the tree. No control measures are needed for this condition.

Leaf crinkle

Leaf crinkle virus causes crinkling of leaves of some citrus but does not seem to reduce cropping. Citrus

Citrus leaf showing leaf crinkle, scale insects, citrus whitefly nymphs (white) and magnesium nutrient deficiency (yellow blotches in leaf colour)

budwood schemes now ensure that virus-free budwood is available for propagators. Because most viruses are transmitted vegetatively by grafting or budding, their incidence has reduced. Crinkling can also be brought about by lack of potassium or by severe restrictions during the growth (expansion) of the leaf.

Leaf drop
Leaves of citrus trees have a life expectancy of around 3–4 years; they will then drop from the tree. Leaf drop can also be caused by a severe drop in temperature caused by frosty conditions or snowfall, by moisture stress or by nutritional stress and, if all leaves are suddenly lost, trees can respond by immediate full flowering. This characteristic has been used in the past to encourage citrus trees to blossom (for their perfume) throughout the summer season, especially trees grown in pots where water supply is easily controlled.

Root-bound potted citrus will become unthrifty, lack new growth and drop some leaves.

Lichen
Lichen is seen as light green, brown or white patches of flat, fluffy or crinkly feathery growth on limbs and branches of many types of fruit trees including citrus. Lichen is often seen on tree trunks or dead twigs and branches. Lichen is the product of a fungus and algae living in symbiosis and does no harm to trees. In fact, if lichen is present it represents a clean, non-polluted environment, so it is a good sign, and no control measures are necessary. See photo, page 92.

Mutation
Variegated coloration of leaves, differing segmented parts on citrus fruit, distorted fruit (excluding distortion caused by mites) are often the result of genetic differentiation expressed in leaves or fruit and are not caused by disease. Mutations may produce new or better fruit or ornamental plants such as variegated lemon trees for use in the home garden. See photo, page 92.

Over-large leaves
Abnormally sized leaves may occur if a tree is cut down severely and the regrowth is strong, or if new scions are grafted into an old tree stump. Large leaves can also indicate excess nitrogen has been given to the tree.

Salt
Excessive salt, whether wind blown, from ground water, or because of too much fertiliser, will cause burning of leaf edges and poor tree health.

In severe cases, shoots will blacken and the tree will die. To avoid salt damage, it is important that irrigation water contains few dissolved salts. Adequate drainage needs to be provided and care should be taken when applying fertilisers to the tree. Where wind-blown salt is a problem, protective shelters can be built or citrus trees can be surrounded by a barrier of salt-tolerant plants.

Split skins

Skin splitting of citrus fruit and subsequent opening of inside flesh, often only on a few fruit on the one tree, is a non-pathogenic condition that affects most citrus species. There are various causes, including: moisture stress; sudden growth spurts within the tree; weather conditions that warm fruit during the day but frost them at night; lack of certain nutrients such as copper and calcium; and the general health of the tree. Disease or insect infestations such as caused by scale may also place stress on fruit skin reducing its ability to flex with growth phases. Ensuring good nutrition, adequate watering, and a calcium supply through regular applications (every few years) of lime around the tree, can help prevent fruit splitting. In alkaline soils, trees can be given gypsum instead of lime.

Sun scorch

Fruit, leaves and bark can be affected by sun scorch. Leaves become burnt and curled, while fruit shows discoloured skin that may go soft. Heat scorch to healthy trees is very rare as citrus trees are, in general, very resistant to hot conditions. In areas with low humidity, however, heat scorch is possible and some shading (such as shade-cloth) will be necessary. Trees under water stress or with exposed roots not kept cool by mulching, may also be subject to heat scorch. If tree foliage is reduced exposing bark, then to protect it bark should be painted with white water-based paint.

Thick skins

Citrus fruit may develop thickened skin, particularly grapefruit and lemon cultivars that tend to develop thick skin as they age. Thick skin can also be associated with excess tree vigour in young trees, regrafted trees or trees that have been given too much nitrogen fertiliser. Seedling-grown trees may also produce thick-skinned fruit.

Some citrus fruit develops thick skin and a 'puffiness' as the skin becomes detached from the flesh. The flesh may become granulated and dry. These are physiological symptoms relating to the age of the fruit, the trees' management regime and to environmental conditions. If trees are managed well and not allowed to become water stressed these problems should not occur. See photo, page 92.

Waterlogging

Most citrus trees have shallow root systems and need watering even during winter, but where there is poor drainage, trees can become waterlogged, a disastrous situation for trees. Symptoms are easy to detect: leaves will droop, show water-soaked areas on the leaf and lose colour. The tree may gradually die. See photo, page 92.

Nematodes, citrus nematode

These microscopic 'worms' (*Tylenchulus semipenetrans*) are not classified as insects, disorders or diseases. They infect most older plantings of citrus, especially in areas with sandy soils. They often prove to be a problem when replanting a citrus grove. The worms are active in the soil, and once females start feeding they become embedded and remain in one place. If huge numbers of nematodes are present on the roots of a citrus plant, it will decline, produce smaller fruit, and may develop leaf drop and twig die-back.

To prevent this problem from occurring, it is wise to select nematode-resistant rootstocks and to make sure that nematode-free trees are purchased. Once nematodes are introduced into an area it is very difficult – almost impossible – to eradicate them all. Fumigation of the soil has been tried and various nematacides (chemical controls), but these are not suitable for home garden use. Building a highly active biological soil regime may help to partially control nematode numbers.

Nutritional deficiencies of citrus

The type of soil that citrus trees are planted in will affect the growth of the tree. Poorly managed soils will have few nutrients to supply to the plant. Heavy clay soils can cause waterlogging of roots and collar rot if no drainage is provided. Waterlogged roots cannot take up nutrients, so the plant can show various nutrient deficiency symptoms on its leaves.

The pH of the soil (its acidity or alkalinity, usually measured from pH3.5 – very acid – to pH10.0 – very alkaline) can cause soil nutrients to become unavailable, making nutrient deficiency symptoms show on leaves and shoots.

This can be seen when fertilisers such as sulphate of ammonia, an acidifying fertiliser, are used to excess: leaves grow large but may yellow as the soil becomes acidified.

Conversely, an out of balance pH can create conditions where some nutrients are in oversupply causing toxicity symptoms to develop, such as burnt leaf edges. Simple to use pH kits are readily available for home gardeners for testing pH levels and should be used regularly.

Acid soils, typical of some Australian sandy soil types, often lack nutrients and will need to be regularly supplied with plant food to grow citrus trees successfully. Very acid soils, by their very nature, prevent plants from absorbing certain nutrients even when they are present in the soil. The main nutrients made less available by severe acidic conditions are nitrogen, phosphorus, potassium, sulphur, calcium, magnesium and molybdenum.

Very alkaline soils act in a similar way, preventing nutrient uptake. Very alkaline soils also affect the availability of nutrients such as iron, nitrogen, manganese, copper and zinc, making them unavailable to plants until the soil has been made more acid.

Some soils lack certain specific nutrients such as copper and zinc. Altering the pH of the soil will have no effect on the availability of these nutrients, so specific nutrients must be added directly to the soil or via foliar sprays.

The ideal pH for most plants is pH 7.0, but the range of pH 6.5 to pH 7.5 is usually safe, allowing the uptake of all essential plant nutrients. If plants are grown to organic standards, using animal manures, compost and mulch, the pH of the soil will gradually become neutral (pH 7.0) or slightly acidic, which is ideal for most citrus plants.

Citrus trees can show symptoms of many of the nutrient deficiencies in their leaf pattern and colour and these can be identified readily. Symptoms of deficiency may be accentuated by high or low pH, in which case the pH should be adjusted before any fertiliser is applied. In other cases, nutrient deficiency symptoms may be brought about because of an oversupply of other plant nutrients; zinc deficiency symptoms are, for example, accentuated by excess phosphate or nitrogen. Other

SEAWEED PRODUCTS TO SUPPLY NUTRIENTS

Although little work has been done with seaweed foliar sprays on citrus, it is worth trying these to partially remedy nutrient deficiencies. Liquid seaweed products such as Maxicrop™, and Vitec™ and Powerfeed™ liquid fish food products, are being used successfully to improve the health of citrus trees and to add some necessary micronutrients. Seaweed extracts contain most elements known to science and can supply them in minute quantities when used as root drenches or foliar sprays. Seaweed as a root drench will improve the formation of fibrous roots, thus increasing fertiliser uptake and preventing some deficiencies. Indirectly, seaweed soil drenches improve biological activity within soils, helping create slightly acid or neutral soil pH ideal for citrus trees. Foliar sprays allow nutrients to be absorbed immediately.

conditions can cause nutrient-deficiency-like symptoms (poor root growth, excess salt, irrigation water with high salinity, waterlogging, tight [compacted] soils, poor drainage, water stress, cold weather, heat stress, chemical pollutants in the soil profile, or weed killers that are sprayed onto or near the trees by accident), so it is worth checking for these.

Citrus trees grown in pots can be more prone to some nutrient deficiencies because every time pots are watered, nutrients are flushed away and need regular replacement. Trees in pots can become root bound, affecting nutrient uptake. Roots can be scorched in heat waves or can strangle the plant. It is important that micronutrients be applied if needed but in very small amounts to avoid over-supply leading to toxicity. Packaged quality micronutrients are readily available to home gardeners.

The main nutrient deficiencies and toxicity symptoms seen on citrus trees

Boron deficiency

Typical boron deficiency showing in citrus fruit flesh

Lack of boron affects flesh of fruit. Individual cells become dry and less juicy and they may have a brownish tinge just under skin next to the dried out flesh. Leaves can become thickened with thickened veins on the top surface, showing splits. Often leaves are brittle and can show a downward curling. Leaf colour is sometimes bronze-yellow and multiple buds may form on twigs giving a bushy appearance. Boron deficiency rarely occurs in heavy soils, but may occur when citrus are grown in deficient or very 'light' sandy soils. Young trees grown in poor soils or poor soil mixes in pots can show symptoms.

Conversely, boron toxicity can occur when too much borax is given to garden plants to correct a boron deficiency or when plants are grown on heavy soils with limestone in the soil profile and subject to severe waterlogging. Toxicity symptoms show as twig dieback, poor growth and when less fruit is produced per tree. Boron toxicity may also produce a burning of leaf edges, some inter-veinal yellowing, and blackish-brown pustules growing on the backs of leaves. Leaves may fall, leaving the petiole (leaf stalk) still attached to the tree. Boron toxicity may occur if too much borax is applied to the tree or pot grown plants.

Animal manures or organic fertiliser mixes containing boron can be added to soil or, alternatively, borax

The main nutrient deficiencies and toxicity symptoms seen on citrus trees

may be applied at a rate of 30–60g per tree and watered in. For large pot grown plants add only one teaspoon of borax per pot. Toxicity symptoms can be fixed by flooding the area to leach away excess boron or by replacing all the soil in pots.

Calcium deficiency

Lack of calcium decreases plant health and vigour. Calcium is necessary to give strength to plant cells. In relation to citrus, calcium deficiency is thought to be one of the factors associated with splitting of citrus fruit.

Calcium is usually readily available in most Australian soils. Many soil types have calcium, limestone or marine deposits in the soil profile and calcium gradually becomes available as these break down. Calcium is easily leached from the soil, though, and needs replacement especially in sandy soils. Applications of lime around citrus trees will adjust soil pH and supply calcium.

This lemon has split as a result of calcium deficiency; it also has scale insects on it, while the leaf under it shows blotches signifying magnesium deficiency

Copper deficiency

Grapefruit splitting that can be aggravated/caused by a copper deficiency

Some areas of Australia have soils naturally deficient in copper. Extremely acid or very alkaline conditions will also aggravate copper deficiency symptoms. Citrus trees produce bunched dark green leaves and short shoots with short internodal sections. Gum pockets may form in laterals or be exuded and some fruit develop roughness and are likely to split. Toxicity with excess copper will cause shoots to wilt and die. These symptoms may occur on old orchard sites where soils can contain high residues of copper from past spray programs.

For copper deficiencies, copper sprays can be applied. Sprays such as Bordeaux (copper sulphate), however, normally sprayed onto citrus trees for disease control, have enough copper to remedy most copper deficiency problems. Excess copper must be leached out of the soil profile by watering heavily. In bad cases, soil in the planting area will need complete replacement.

The main nutrient deficiencies and toxicity symptoms seen on citrus trees

Iron deficiency

Most Australian soils contain adequate iron for plant growth, but iron deficiency is one of the most common leaf symptoms seen on citrus. It can occur in highly alkaline soils, or be caused by too frequent applications of (very alkaline) wood ash, runoff from builders' lime around building sites, or over application of garden lime to the soil around a tree. Iron deficiency symptoms can show when plants are grown in cold wet or waterlogged soils, or when over-irrigation leaches out nutrients (especially in pots).

Lisbon lemon leaf showing typical iron deficiency symptoms (Iron chlorosis)

Extremely alkaline soils produce 'iron chlorosis', where leaves lack normal vivid dark green colour. Green leaves can turn creamy white to white but some veins may remain green. Usually leaves are of normal size because other nutrients are not being inhibited by the high alkalinity.

Adjusting pH by adding acidic materials such as compost, peat or chemical acid fertilisers such as sulphate of ammonia is the main treatment. For a 'quick fix', iron chelate mixed with water can be sprayed or watered onto the foliage to turn leaves green. Toxicity symptoms are not usually seen on plants grown in Australia.

Magnesium deficiency

Lisbon lemon leaf showing typical magnesium deficiency symptoms

Magnesium deficiencies can occur when citrus are grown in severely acidic soils or when the soils naturally lack magnesium. Poorly managed plants grown in pots can also show symptoms. An oversupply of the plant nutrient potassium will aggravate magnesium deficiency. Symptoms first occur as yellow blotches in between the veins of leaves; these become more pronounced as deficiency progresses and leaves may turn yellow with only a few patches of green at the leaf edge, tip and stalk end. Leaves sometimes have a colour segmentaion of yellow and green, the green forming a broad inverted 'v' at the base of the leaf. Some leaves may drop from the tree and foliage and twigs look sickly.

The main nutrient deficiencies and toxicity symptoms seen on citrus trees

To treat organically, add Epsom salts (magnesium sulphate) to soil. Magnesium foliar sprays can also be applied to leaves whenever a growth flush is evident. Dolomitic lime can be used in acid soils.

Toxicity symptoms caused by magnesium do not usually occur.

Manganese deficiency

Manganese deficiencies can occur on normal sized leaves when citrus trees are grown in very acid or highly alkaline soils, and shows as a gradual chlorotic yellowing between the veins of the leaf. Magnesium deficiency produces blotchy yellowing whereas manganese deficiency produces yellowing of the whole leaf area except veins, which remain green. Treatment is to correct pH or use foliar sprays of manganese sulphate on spring growth. Note that zinc and manganese sprays can be applied at the same time.

Lisbon lemon leaf showing typical manganese deficiency symptoms

Toxicity from excess manganese only shows in pot culture and gives the leaves a chlorotic effect not unlike iron deficiency.

Nitrogen deficiency

Lisbon lemon leaf showing typical nitrogen deficiencies symptoms

Lack of nitrogen will cause small pale green to yellow leaves, a lack of healthy growth, developing fruit will fall and remaining fruit will be small. Treat by adding organic manures such as fowl manure or other fertilisers with high nitrogen content.

An oversupply of nitrogen will produce large green leaves, excessive growth and less fruit. Fruit skins can also become very thickened. Toxic amounts of nitrogen especially in the form of nitrogen salts will kill plants.

The main nutrient deficiencies and toxicity symptoms seen on citrus trees

Phosphorus deficiency

Phosphorus is essential for plant growth and can be naturally deficient in some Australian soils. If prescribed fertiliser applications are given, this deficiency will rarely occur. Symptoms may occur with poorly managed pot-grown plants. Symptoms include bronzing of new leaves, dullness of old leaf colour, leaf drop and thinning of foliage associated with poor growth of the tree shoots; fruit may become swollen with bumpy skins and have a low juice content. To remedy, apply fertiliser containing phosphorus, organic manures or rock phosphorus dust.

Toxicity symptoms would likely show leaf margin scorch but this does not seem to occur with citrus grown in Australia.

Tahitian lime tree foliage showing poor growth and die-back that can be caused by insect infestation, drought, twig fungal diseases, poor nutrition and phosphorus deficiency

Potassium deficiency

Lisbon lemon foliage showing twisting and yellowing of foliage typical of potassium deficiency symptoms

Citrus rarely suffer potassium deficiency but it may occur in pot-grown trees subject to nutrient leaching, or in trees grown in poor, extremely acid sandy soils. The most common symptom is degeneration of leaf material in small patches with some yellowing followed by death of tissue giving the effect of leaf scorch as patches join together. Leaf scorch can occur within or around edges of leaves. In soils deficient in potassium, there may be leaf crinkling, twisting of the leaves and yellowing of mature leaves. Leaf drop after flowering and some die-back of twigs will occur. In poor very acidic, sandy soils the potassium may be in the soil profile but not available to citrus plants. In severe cases the new shoots show a curling of stems on new shoots. At a pH of 4.0, or below, plants can use almost none of the available potassium. Adjusting soil pH and giving usual

The main nutrient deficiencies and toxicity symptoms seen on citrus trees

applications of animal manures or mixed fertiliser containing potassium will generally supply sufficient potassium to correct deficiencies.

Symptoms of potassium excess show spotting of leaves and leaf discolouration but this does not usually occur with home-grown citrus trees.

Sulphur deficiency

The importance of sulphur has been upgraded. While still officially classed as a micronutrient, it is now regarded as one of the major nutrients for plant growth and very necessary for citrus plants. Sulphur is often naturally available from rainwater, from soils containing various sulphates and from biologically active organic composts so that trees grown in the open will rarely show a deficiency. Symptoms may occur when sterile soils are used to grow citrus in pots or where soils become waterlogged and no nutrient uptake can occur.

Lisbon lemon tree foliage showing bronzing of new growth and yellowing of new leaves growth, typical of sulphur deficiency symptoms

Symptoms show, particularly on young leaves and shoots, as an overall yellowing although adjacent older leaves are still green but not vividly so. Trees will lack vigour and there will be some leaf drop with dead twigs evident over time. Trees may have over-abundant, weakened flowers and any fruit that forms can be misshapen with thick skins, little juice and with dull skin colour when mature. To prevent sulphur deficiency, firstly check soil pH. Materials such as iron sulphate, potassium sulphate, gypsum, or other sulphate nutrient fertilisers will supply additional sulphur.

Excess sulphur shows as a typical salt-related leaf burn, leaf drop and eventually, defoliation of the tree. To correct this, the soil profile will have to be leached thoroughly with water to wash away the nutrient salts.

The main nutrient deficiencies and toxicity symptoms seen on citrus trees

Zinc deficiency

Grapefruit foliage showing zinc deficiency symptoms

Symptoms can occur on leaves of citrus trees growing in very acid or highly alkaline soils and in soils that do not contain any zinc. Symptoms are distinct in that mature leaves show a severe bright yellow mottling from the leaf tip downwards, with the green around the midrib forming a inverted 'V' with its tip towards the leaf end. As the deficiency progresses, young leaves may show almost complete yellowing with a few veins still green and the tree may become stunted and bushy.

To treat, adjust pH if necessary, or use fertilisers containing zinc. For quick results, apply zinc sulphate foliage sprays. The best time to apply foliar fertilisers is just as a new growth flush is occurring. Continued applications of zinc may be needed.

Soils high in zinc such as those found on some industrial sites converted to residential use, will make citrus trees grow poorly and cause yellow leaves. Soils with extremely high zinc content will have to be replaced to ensure good growth of citrus.

COLD WET WINTERS

When citrus trees growing in cooler climates experience cold wet conditions during the winter period they can show 'leaf yellows' – a yellowing of the leaves on the tree. This is caused by cold weather conditions and the cool soil temperature. A similar effect is observed with leaves of the passion fruit vine grown under the same conditions. When the soil warms again during spring/ summer, the leaves of the affected plants will again turn their normal green.

A–Z of organic pest and disease control

There are an increasing number of organic methods and materials for controlling pests and diseases, and the following should be used as a general guide only.

Aeration – pruning plants to allow air flow through them, and to enable light to penetrate the tree to prevent the build-up of pests and diseases.

Baking soda – often used in diluted form to control diseases.

Basal watering – watering around the base of plants only: a water saving method that also prevents water getting onto the stem or foliage of plants, thus helping prevent fungal and bacterial diseases.

Beat-A-Bug™ – a product containing pyrethrum, garlic and hot chilli mixed together; useful for controlling insects.

Beer traps – stale beer in slippery containers attracts slugs and snails.

Biological controls such as parasitic insects, see *Insect predators* below.

Birds – attracting insectivorous birds such as honeyeaters, and some native animals (e.g. snakes, lizards, antechinus).

Bordeaux – mixture of copper sulphate and lime that mainly controls bacterial diseases.

Care – with handling, sorting, clipping of fruit.

Compost – aids the health of trees by adding some nutrients, humus, organic particles to improve soil structure and many organisms such as worms, fungi, bacteria and yeasts that help increase biological activity and soil fertility.

Copper tape – has been used around plants to stop snails and slugs.

Crop rotation – planting different plant family groups in individual garden plots each year for four years in a rotation system to avoid build-up of pests and diseases specific to one family group.

Derris Dust – dust made from plant material; controls several insect species.

Dipel™ – a biological bacterial product that attacks larvae (grubs) of several insect species including cabbage white butterfly and light brown apple moth.

Eco-oil™ – an organic, oil-based miticide/insecticide, useful for aphids, scale and citrus leaf miner control.

Enviromat™ **mulch collars** – matting used as collars around plants for weed control purposes.

Fish food – emulsion and foliar sprays for tree nutrition.

Frost protection – by planting wind breaks, using tree guards or fencing.

Fruit fly traps – attract and kill fruit fly adults; used as an indicator of fruit fly populations.

Hydronurture™ igloos – water-filled cellular ribbed igloos that can be used to protect young plants and trees to give extra warmth, which boosts plant growth, especially during cool weather periods.

Hygiene – especially when budding and grafting, or handling fruit.

Insect predators – mites, eelworms, wasps, ladybirds and parasitic wasps (*Comperiella bifasciata* and two *Aphytis* spp. for the control of Red Scale, are included in this group). These can be purchased from suppliers and released into the garden/orchard to control pest insects.

Jiffy pots™ – compressed peat moss plugs provide disease free growing media for seed germination and plant propagation.

Lime dust/powder – used for making Bordeaux mix and Bordeaux paste; lime wash used to control some diseases and the dust is used to control pear and cherry slug populations (see also Glossary).

Lime sulphur – a corrosive, smelly liquid used as an organic fungicide spray to control some bacterial and fungal diseases.

Maxicrop™ – see *Seaweed, liquid* below.

Milk – diluted milk or whey used as a spray to control powdery mildew diseases.

Mineral oil – used to control insects and to kill insect and mite eggs.

Moisture – adequate moisture at all times.

Mulch and mulch mowing – used to build up the soil organic matter and improve soils, keep plant roots cool and help control weeds.

Multiguard™ Slug and Snail Pellets – an organically registered material used for slug and snail control.

Natrakelp™ – see *Seaweed, liquid* below.

Neem – plant extract that acts on insects, making them virtually starve to death.

Netting – plant or orchard covering used to prevent predation by animals and birds.

Orange oils (citrus oils) – used to control some insects; they kill their eggs.

Paper bags – placed over fruit before it ripens, to prevent bird, flying fox and fruit fly attack; can also be used to protect fruit from other insects.

Pestoil™ – organically registered oil product useful for controlling many insects such as scale, citrus leaf miner and aphids.

Pheromone traps – these sticky traps or lures attract and capture male insects with a pheromone; as a result male insects are prevented from breeding.

Powerfeed™ – a liquid fish waste/compost mix used through the irrigation system or by watering into the soil around plants for the supply of nutrients and as a crop enhancer. This mix increases root mass in many food plants. Keeping plants healthy aids pest and disease control.

Pruning – opening up trees to allow air flow and light into the centre of the foliage area to help prevent the build-up of pests and diseases, or to remove diseased branches or twigs (see Chapter 5).

Pyrethrum – broad spectrum spray that will kill all insects but must be used with care; read product labels and follow instructions.

Quassia chips – tree chips that are steeped in boiling water to extract a bitter liquid that is then diluted and sprayed onto foliage to deter possums.

Removing – infected fruit.

Seaweed, liquid – various brands available; used as soil drenches or as foliar sprays; provides minute quantities of nutrients for plants, encourages fibrous root systems and seems to help the build-up of pest and disease resistance in some plants.

Soapy water – smothers scale insects and their eggs and kills some other insect species.

Soil – preparation, testing.

Sticky traps – usually hanging oblong, round or cylindrical shaped material with very sticky glue on the surface to trap flying insects such as whitefly.

Tea tree oil – used as an antiseptic, cleansing agent and for the control of some diseases.

Vitec™ **Fish Food Emulsion** – fertiliser used as a foliage spray especially on citrus trees; it can be used as a root drench and seems to help build up resistance to some plant diseases.

Weed control

Worms – encourage worms. These essential inhabitants of healthy soils help improve soil structure, add nutrients to the soil, thus aiding the health of trees. Worms aerate the soil allowing plant roots, bacteria, fungi and other beneficial organisms to flourish and, as well, they help incorporate organic matter into soils.

CHAPTER 7

harvesting and storing citrus fruit

Citrus fruit is usually harvested when the skin colour becomes orange or yellow or, for oranges and other sweet citrus, when sugar content is right. It must be noted that citrus fruit grown in tropical to semi-tropical areas may fail to develop fully coloured skin and will remain a greenish colour.

Fruit such as Valencia oranges and some other citrus cultivars have green skin before ripening and may regreen after becoming ripe and fully coloured.

Most citrus fruit can be left on the tree for some time, in some cases for a month or two, after ripening without too much decrease in the quality of the fruit. If left on the tree too long, however, most citrus swell up with moisture, the skin becomes loose and 'puffy' and the taste deteriorates. Mandarins particularly should not be left on the tree for too long after ripening, as the flesh dries out and skins separate from the flesh.

These problems aside, home gardeners can enjoy fresh citrus fruit picked from their own trees for a long period especially if they have more than one cultivar or several species ripening at different times of the year. Some citrus, such as Valencia oranges, will have ripening fruit on the tree at the same time as new flowers.

Because individual tastes vary, it is difficult to give a definite indicator of when sugar content is at the right point for picking. Most citrus fruit begins as very hard dark green fruit. As it reaches mature size, the skin starts to turn from green to orange or yellow. It may still be months before fruit reaches optimum sugar content and flavour, as the acids within the fruit decrease and sugars become more prominent. If one fruit picked from the tree tastes good to you, then generally the fruit will be ripe enough to pick.

Commercial growers pick fruit to a legal maturity standard which is based on TSS (total soluble solids) and TTA (total titratable acidity) of the juice. Fruit picked at this stage will probably taste acidic to most people but the acidity lessens as the picked fruit ages.

Different species and cultivars can be grown to supply year-round fresh fruit. Navel oranges, for example, ripen during autumn–winter, mandarin cultivars ripen over a long period from late spring to mid-winter, Valencia oranges ripen from late spring to autumn, while grapefruit are usually available during late summer to winter. Tangors are a late winter to early summer fruit, while Tangelo 'Seminole' ripens early spring to mid-summer, and cumquats are available in winter. Limes ripen throughout the summer months, and lemons and limes are available commercially nearly all year round. Fruit maturity will vary in different climate zones and will also depend upon factors such as the type of rootstock used.

Harvesting citrus

To pick, citrus fruit is usually twisted and pulled from the tree. Mandarins, however, should not be picked this way as a small 'plug' of skin may pull out, leaving the green button hanging on the tree. Exposed flesh is then open for infection, and storage capacity much reduced. Clipping fruit from the tree is the preferred option for mandarins.

Harvesting Imperial mandarins with secateurs and cotton gloved hands.

Citrus should not be picked, especially for storage, when the fruit is wet with dew or rain. This makes the skin cells swell and fruit is easily bruised or damaged by rubbing. Fruit will not store for long if too much moisture is present.

Citrus fruit must be handled with care. Wearing cotton gloves is a good idea but, if picking with bare hands, make sure fingernails are short to prevent them injuring citrus skin and providing sites for blue or green mould infections. Gently lower fruit into boxes or containers. Do not place containers in direct sunlight where fruit may be sunburned. Pick during the cooler part of the day if possible.

Storage of citrus

Most fresh citrus have a limited storage life, so storage methods to extend citrus availability are important.

Once picked and if not stored appropriately, citrus fruit dehydrates, the skins get thinner and there can be an increase of juice by up to 30%.

To judge the condition of fruit for sale in the market, look at the 'button' (calyx) at the stem end of the fruit. This is a waxy light-green when picked; after gassing and storage it may turn brown-yellow or black or fall off completely. Fruit at its optimum that has not been stored too long will have firm fresh green buttons. To test for juice content, weigh fruit in your hand and 'feel' the difference between heavy and light fruit. Once home, another way to test for juiciness is to place fruit in water: fruit 'heavy with juice' will sink but fruit with little juice will not.

Commercial juice manufacturers take advantage by 'curing' fruit before processing to obtain more juice. The amount of pith decreases, fruit skin becomes thin, hard, and leathery, and the skin changes colour. While such fruit is not very appealing to look at, the taste of the juice is really something.

Citrus fruit can be stored for a short time in plastic bags in the crisper section of the refrigerator, or in open fruit bowls if the fruit is to be eaten very soon.

Cool storage of citrus fruit can about double life expectancy compared to fruit stored at room temperature. Mandarins and sweet oranges are best stored at a temperature of 7°C; grapefruit, lemon and limes at a temperature of 10°C. To extend their life, fruit can be wrapped individually in wax paper, and stored in well-aerated containers in a cool dark place. The skins may become hardened and tough and not look inviting but the juice will be excellent to taste. Fruits must be separate to avoid fungal infections such as blue mould spreading from fruit to fruit.

Citrus peel can be candied, citrus juice can be frozen and citrus slices can be dried for longer storage. One of my favourites is dried orange slices dipped in dark chocolate.

Dried citrus fruit slices in sealed jars

Jar of cumquats prepared for preserving

Commercially, citrus fruit is cool-stored for up to five months at 5–10°C (depending on species) and 98% humidity. Fruit changes over time to become less acidic to the taste, has thinner skin and more juice. The fruit is usually washed and brushed to remove dust, dirt and scale insects, and then dipped in fungicides to prevent rots. Fruit is often waxed to make them look fresh and to aid storage, and they are gassed with ethylene to promote skin colouring. At this stage, fruit may still taste a little acidic, but will continue ripening slowly even while stored. Fruit picked early in the season will still have a considerable acid content even though they may look like fully ripened fruit. Conversely, waxed and gassed fruit kept in storage for three months or more before sale will taste very sweet, but compared to sun-ripened fruit picked straight from your own tree, will have a distinctive taste.

It is a fallacy that citrus skins and discarded pulp cannot be composted because they do not break down. With some help, citrus waste can be composted. Bulky or deep layers of citrus waste will not readily break down, but if the layers are kept thin (about 1–2cm) and mixed thoroughly with other materials, composting will occur. Drying citrus skins before placing them in the compost, or shredding the citrus waste finely and mixing it with other compost materials, will also help ensure composting success.

Lolly made from cumquat fruit, China

Preserved citrus, Guangzou market, China

LEMON FACTS

Lemons provide juice for lemon meringue pies, lemonade drinks, candied peel, slices for alcoholic drinks, cooking, flavouring or for medicinal uses such as honey and lemon juice with hot water to 'fix' a cold. Fish and lemon are inseparable.

Lemon fruit can be made to 'give up' more juice if they are placed into a microwave apparatus for a few seconds to make them juicier.

In my naïve days of being introduced to alcoholic drinks; I remember consuming several small glasses of tequila, and the 'in' way to drink it was to squeeze lemon juice, sprinkle salt onto your wrist, lick this, then consume the liquid in one gulp; I remember the gulps but not the rest of the night so don't try this if you are out for the night.

Products obtained from the lemon, including oil, pectin and citric acid, and they are used in a whole range of citrus products such as cleansers.

Lemons have many vitamins particularly vitamin C and vitamin A, and lemon products are used for the care of hands and teeth, for facial applications, and to improve hair health.

APPROXIMATE AMOUNTS OF FRUIT JUICE AND PEEL

1 orange gives 60–90 mL juice and 3 tsps peel

1 lemon gives 40–60 mL juice and 2–3 tsps peel

1 lime gives 30–40 mL juice and 1–2 tsps peel

1 grapefruit gives 155–185 mL juice and 3–4 tsps peel

1 mandarin gives 60–90 mL juice and 1–2 tsps peel

TAMARI AND ORANGE JUICE SALAD DRESSING

To the juice of one orange add an equal quantity of Tamari. Taste for sweetness and add brown sugar or honey if desired. Add a couple of drops of sesame oil and a touch of chili if liked. Good on green salad and particularly on avocados.

Lime juice can be substituted for orange juice but add more honey or sugar.

GRAVY'S LEMON DELICIOUS PUDDING

Cream 60g butter and $3/4$ cup sugar. Add grated rind of 1 lemon. Sift $1/2$ cup SR flour and add to mixture in two lots beating well after each addition.

Separate 2 eggs. Beat egg yolks and add 1 cup milk. Add to mixture in small amounts alternately with 3 tbsps of lemon juice. Lastly, fold in stiffly beaten egg whites.

Pour into greased pie dish, and stand dish in a baking dish containing cold water. Bake in moderate oven for about 40 minutes until top is firm and lightly browned.

NORMA'S MARMALADE

Norma Campbell, a well-known and much loved Bruny Islander, sold marmalade and jams at markets on Bruny Island for many years. I particularly liked her marmalades. Norma was kind enough to give me some of her recipes for this book.

Wash 8 oranges and 8 lemons and slice very thinly, removing pips and excess pith and putting these in a muslin bag as a pectin source. Cover fruit with water allowing 2.5 L of water for each kilogram of sliced fruit. Let stand for 24 hours.

Grease pan and, if you have old two-shilling pieces, put 3 or 4 of these in the bottom of the pan. Do not use modern 20 cent pieces. This step is not essential but helps prevent sticking.

Weigh fruit and water together, then warm on stove in a large preserving pan.

For each 500 g of fruit/water mix allow 2.5 cups sugar. For a less sweet marmalade allow $3/4$ cup of sugar to each 1 cup combined fruit/water. Warm sugar in oven to approximately the same temperature as the fruit mix, then stir sugar into fruit mix stirring until completely dissolved without boiling. Add pith and pips.

Boil all together very quickly until the orange pieces are transparent and setting point is reached (see Glossary).

Remove any scum with a perforated spoon.

Remove bag of pith and pips and bottle into clean, sterilised jars, stirring as little as possible. Seal with jam covers.

NORMA CAMPBELL'S BRANDIED MANDARIN MARMALADE

Wash and cut twelve smooth-skinned mandarins in half, slice finely and put in preserving pan. Put seed in a muslin or cheesecloth bag and add to pan. Cover with water and stand 24 hours. Boil gently until fruit is tender (approximately 45 minutes). Squeeze and remove muslin bag. Measure combined fruit and liquid, return to pan and add $^3/_4$ cup warmed sugar to 1 cup of mixture. Stir gently until dissolved. Add juice of 3 lemons. Boil quickly until setting point (see Glossary) is reached. Remove from heat and gently stir in 1 cup of brandy, avoiding air bubbles. Return to heat and boil briskly for 5 minutes, stirring gently. Remove scum with a perforated spoon. Allow to cool for 10–15 minutes and bottle.

NORMA'S SEVILLE ORANGE MARMALADE

Clean and finely slice 1kg Seville oranges and 250g lemons, removing and setting aside excess pith and seed. Cover fruit with $1^1/_4$ L water and leave for 24 hours.

Boil fruit and water mix slowly for 45 minutes (or until slices of fruit soft) then let stand for another 24 hours. Test for pectin (see Glossary).

Boil seed and pith with a little water until soft and nearly all water gone; strain and add to fruit.

Measure, and heat without boiling. Add $^3/_4$ cup warmed sugar for every cup fruit. Stir in thoroughly, then boil rapidly and test frequently after first half hour. Do not stir.

When setting point (see Glossary) reached, remove any scum and leave for 10 minutes before gently pouring into sterilised jars. Cover, seal and label.

NORMA'S GRAPEFRUIT AND GINGER MARMALADE

Clean and finely slice 1kg grapefruit and 2 lemons, setting aside pith and seeds. Boil seed and pith in a little water, strain and add to water to make 8 cups. Cover fruit with this mix and stand for 12 hours.

Simmer gently for $1^1/_2$–2 hours until fruit is tender.

Test for pectin (see Glossary). If not satisfactory add grated lemon rind or boil longer.

Measure pulp, add $^3/_4$ cup warmed sugar for every 1 cup pulp. Add 200 g finely chopped glacé ginger and 1 tsp ground ginger (for stronger ginger flavour). Heat, gently stirring until all sugar dissolved. Heat rapidly until setting point reached (see Glossary). Remove any scum. Pour gently into warm, sterilised jars. Seal and label.

NORMA'S LEMON MARMALADE

Wash 1kg lemons (or limes). Squeeze out juice and set aside. Remove pips and excess pith and place in muslin bag. Finely slice skins. Soak skins in 13 cups liquid (juice plus cold water) with pips and pith in bag.

Slowly bring to boil and simmer until slices of lemon are tender or until pectin test (see Glossary) is satisfactory.

Remove and squeeze out muslin bag. Add $^3/_4$ cup warmed sugar for every cup of pulp, stirring until sugar is completely dissolved. Boil rapidly until setting point is reached (see Glossary). Skim off scum with perforated spoon, cool slightly and pour into sterilised jars. Seal and label.

Ginger can be added in the same way as for grapefruit and ginger marmalade.

NORMA'S THREE-FRUITS MARMALADE

Wash and finely slice 2 oranges, 2 lemons, 2 grapefruit, putting excess pith and seeds into muslin bag. Cover with 10 cups water and let stand overnight.

Slowly bring to simmering point and let simmer for 45 minutes. Remove and squeeze out muslin bag. Add $^3/_4$ cup warmed sugar for every 1 cup pulp (flesh), and stir until sugar is dissolved.

Boil rapidly until setting point (see Glossary) is reached. Cool slightly, bottle, seal and label.

NORMA'S WHISKY MARMALADE

Wash and finely slice 3 oranges and 2 lemons, putting excess pith and seeds into muslin bag. Grate 6 small carrots. Soak all overnight in 3 parts water to 1 part fruit mix.

Gently bring to boil and simmer gently for approximately 1 hour.

Remove and squeeze out muslin bag. Stir in and dissolve $^3/_4$ cup warmed sugar for each 1 cup fruit mix. Boil rapidly until setting point (see Glossary) reached. Add 1 cup whisky and boil for a further 3 minutes. Cool slightly, bottle, cover and label.

LAURIE'S BAKED RHUBARB WITH ORANGE

Remove leaves from rhubarb, wash well and cut into 1cm pieces. Place in glass baking dish, sprinkling liberally with brown sugar and grated orange rind. Dot surface with butter and bake in moderate oven until rhubarb is well cooked and surface is slightly caramelised. Serve with cream or icecream.

NORMA'S CUMQUAT MARMALADE

Wash and finely slice 1kg cumquats. Remove seed and put in 1 cup of water and stand overnight. Cover cumquats with 6 cups water adding grated rind of two lemons; stand overnight. Drain seeds and discard, and add liquid to cumquat mix.

Slowly bring to boil. Simmer 30–45 minutes or until cumquat slices are tender. Stir in and dissolve $3/4$ cup warmed sugar for each 1 cup fruit mix. Boil rapidly 15–25 minutes or until setting point (see Glossary) reached.

Optional: add 1 tablespoon grated fresh ginger. Stand 10 minutes. Test for setting. Bottle, seal and label.

LAURIE'S BREAD AND BUTTER PUDDING

Remove crusts from 10–12 slices of white bread and brush with melted butter and good marmalade. Cover the bottom of a large, flat, glass baking dish with the bread slices, marmalade side up. Sprinkle generously with currants and mixed citrus peel and 1 tbsp of caster sugar. Repeat, finishing with a bread layer.

Beat 4 eggs and 2 egg yolks together. Heat $2^{1}/_{2}$ cups of milk with 2 tbsp caster sugar, and 1 tsp vanilla essence until just warm. Slowly beat in eggs. Strain and pour into baking dish over bread slices. Sprinkle with additional light dusting of caster sugar. Bake for approximately $3/4$ to 1 hour until pudding is brown and puffy with crusty edges. Serve with cream. Serves 6.

LAURIE'S FRUIT COMPOTE

In a heavy-based saucepan, put a generous amount of dried fruit of different kinds (dried apple slices, dried apricots, sultanas, raisins, dried peaches or pears) together with 2 oranges cut into 8–10 segments, 1 stick of cinnamon, and brown sugar to taste. Just cover with orange juice and simmer for 30 minutes. Add as many prunes and dates as desired, add extra brown sugar if desired and simmer for further 15 minutes.

Can be served hot as dessert with cream or icecream but best with muesli for breakfast. Keeps well in fridge.

ORANGE AND ONION SALAD (With thanks to Gwen McCrory)

On a bed of lettuce, arrange thin slices of orange (peel on if very fresh) and thin slices of salad onion. Serve drizzled with vinaigrette dressing or balsamic vinegar.

resources

Organic certification and advice

Australian Certified Organic (ACO), PO Box 530, Chermside, Qld 4032 T 07 3350 5706; www.australianorganic.com.au

Australian Quarantine Inspection Service (AQIS), GPO Box 858, Canberra, ACT 2601 T 02 6272 3933; freecall 1800 020 504; www.aqis.gov.au

Bio-Dynamic Research Institute (BDRI), Main Road, Powelltown Vic. 3797 T/F 03 5966 7333

International Federation of Organic Agriculture Movements (IFOAM) www.ifoam.org

National Assoc. for Sustainable Agriculture (NASAA), PO Box 768, Stirling, SA 5152 T 08 8370 8455; F 08 8370 8381; www.nasaa.com.au

Organic Federation Australia, PO Box 166 Oakleigh South, Vic. 3167 Peak organic industry body in Australia; www.ofa.org.au

Organic Growers of Australia (OGA), PO Box 6171, South Lismore, NSW 2480 T 02 6622 0100; www.organicgrowers.org.au

Organic Food Chain, PO Box 2390, Toowoomba, Qld. 4350. T 07 4637 2600; F 07 4696 7689; www.organicfoodchain.com.au

Organic Future Inc., PO Box 6, Oakleigh South, Vic 3167 Peak organic consumer body in Australia. T 03 8503 7589; www.organicfuture.org

Safe Food Queensland (SFQ), PO Box 400, Spring Hill, Qld 4004 T 1800 300 815

Tasmanian Organic producers (TOP), PO Box 434, Mobray Heights, TAS 7054 T 03 6383 4039

Citrus organisations and resources

Australian Citrus Growers Inc., PO Box 5091, Mildura Vic 3502 Peak industry body in Australia with comprehensive links to other organisations. T 03 5023 6333; www.austcitrus.org.au

Australian Citrus Industry Council (ACIC), 140 Arthur Street, North Sydney, NSW. 2060. T 02 9458 7416.

Sunraysia Nursery and Garden Centre, Sturt Highway, Gol Gol, NSW 2738 T 03 5024 8502; F 03 5024 8551; www.sunraysianurseries.com.au

Useful websites

There are too many to list citrus websites. A search from these will give lots of links, including recipes.

Asian Food Information Centre (AFIC) (non-profit, Singapore) www.afic.org

Association of Societies for Growing Australian Plants (ASGAP) www.farrer.riv.csu.edu.au/ASGAP

Australian New Crops newsletter. www.newcrops.uq.edu.au/newslett

Some other useful organisations

Australian National Botanic Gardens,
   GPO Box 1777, Canberra, ACT 2601
   T 02 6250 9450; www.anbg.gov.au
Bio-Dynamic Agriculture Australia Inc.,
   PO Box 54, Bellingen, NSW 2454
   T 02 6655 0566; F 02 6655 0565;
   www.biodynamics.net.au
Biological Farmers of Australia (BFA),
   PO Box 530, Chermside, Qld. 4350
   T 07 3350 5716; www.bfa.com.au
Henry Doubleday Research Association,
   816 Comleroy Road, Kurrajong, NSW 2758.
   T 02 4576 1220; www.hdra.asn.au
Holmgren Design Services & Melliodora
   (Hepburn Permaculture Gardens),
   16 Fourteenth Street, Hepburn, Vic. 3461.
   T 03 5348 3636; www.holmgren.com.au
Horticulture Australia, Level 1 50 Carrington
   Street, Sydney, NSW 2000
   T 02 9418 2200; www.horticulture.com.au
Permaculture International Ltd., PO Box 219,
   Nimbin, NSW 2480
   www.nor.com.au/environment/perma
Permaculture Research Institute
   www.permaculture.org.au
SEED (Sustainability, Education and Ecological
   Design) International, 50 Crystal Waters,
   Kilcoy Lane, Conondale, Qld 4552
   T/F 07 5494 4833;
   www.permaculture.au.com
South Australian Research and Development
   Institute, GPO Box 397, Adelaide,
   SA 5001. T 08 8303 9400;
   www.sardi.sa.gov.au
Southern Bushfoods Association, 48 Outlook
   Road, Mt Waverly, Vic 3401
   http://home.vicnet.net.au/~bushfood/
Sydney Postharvest Laboratory, CSIRO Food
   Science Australia Building, 11 Julius Avenue,
   Riverside Corporate Park, Delhi Road,
   North Ryde, Sydney, NSW 2113
   (PO Box, 62 North Ryde, NSW 1670).
   T 02 9490 8443; F 02 9490 8499;
   www.postharvest.com.au
WWOOF (Willing Workers on Organic
   Farms), Buchan, Vic. 3885
   T 03 5155 0218; www.wwoof.com.au

Publications

Acres Australia, Freepost1, PO Box 27,
   Eumundi, Qld 4562
   T toll free 1800 801 467;
   www.acresaustralia.com.au
Australian Horticulture, Rural Press Magazines,
   PO Box 254, Moonee Ponds, Vic. 3039.
   T 03 9287 0900; F 03 9370 5622;
   www.ruralpress.com
*Australian Subtropical Fruit and Nut Tree
   Catalogu*e, available from Daley's Fruit Tree
   Nursery, PO Box 154, Kyogle, NSW 2474.
   T 02 66 321 441, www.daleysfruit.com.au
Earth Garden, PO Box 2, Trentham, Vic. 3458.
   F 03 5424 1743; www.earthgarden.com.au
Gardening Australia, GPO Box 9994,
   Melbourne, Vic 3001
   T 9524 2875; www.abc.net.au/gardening
Good Fruit and Vegetables, Rural Press, PO Box
   254, Moonee Ponds, Vic. 3039
   T 03 9287 0900; F 03 9370 5622;
   www.ruralpress.com
Green Connections, PO Box 793,
   Castlemaine, Vic. 3450
   T 03 5470 5040; F 03 5470 6947;
   www.greenconnections.com
Greenworld, Glenvale Publications, PO Box 50,
   Mt Waverly, Vic. 3401
   T 03 9544 2233; www.greenworldmag.au
Landline, GPO Box 9994, Brisbane, Qld 4001
   www.abc.net.au/gardening
The Organic Gardener, PO Box 1067,
   Lismore, NSW 2480
   T 1300 656 933

references

1 Citrus: Origins, scope and spread

[1] Janson in *Pomona's Harvest*, p. 27.
[2] Unusual citrus fruit such as the Fingered Form citron (Buddha's Hand).
[3] *Pomona's Harvest*, p. 113.
[4] Ibid.
[5] *Fungus Diseases of Citrus Trees in Australia, and Their Treatment*, 1899, p. 5.
[6] Early planting of citrus trees placed them about 8m (24 feet) apart, giving approximately 250–270 trees per hectare (100–108 trees to the acre).
[7] McEwin in his *The Fruitgrower's Handbook*, 1913, p. 119.
[8] Ibid., p. 118.
[9] *Horticulture Australia*, p. 335.
[10] Virus free buds are produced by placing selected trees infected with virus into special growth cabinets that are heated constantly to 38°C. The shoot tips that are collected and used grow fast and are usually free of viruses such as Exocortis, that are present in shoots lower down on the plant.
[11] It is interesting to note that prepared citrus juice concentrate has to have 'real' orange juice and peel products added back to the juice after processing, to obtain the 'typical' citrus juice flavour. Fresh orange juice has much more 'true' citrus flavour than the juice concentrate preparation/juice mixes.

2 Citrus species and cultivars

[12] McEwin, H. *The Fruitgrower's Handbook*, 1913. Commonwealth of Australia, p. 119.
[13] Webber, H. J. & Batchelor, L. D. (eds) 1948, *The Citrus Industry*, vol. 1, University of California Press, Berkeley Ca., pp. 530–1.
[14] William Saunders, Superintendent of Gardens and Grounds for the Department of Agriculture.
[15] Tim Herrmann, Manager Auscitrus, pers. comm. 2005 International Plant Propagators' Society (IPPS) Conference, 21–24 April 2005, Mildura, Vic.
[16] Auscitrus 2005 sales figures:
2075 budwood trees containing 108 citrus varieties.
664 seed trees of 108 varieties
Sales 700,000–800,000 buds
600–700kg of seed (approximately 23 million seeds).
[17] Note the approximate plant height and width are shown as a guide only.
[18] William Watson 1936, *The Gardener's Assistant*, vol. 3, The Gresham Publishing Company, London, 1936, pp. 61–5.
[19] Robert Fortune (1812–1880) was first sent to China in 1843 by the Royal Horticultural Society. He remained there for three years and made subsequent visits, the last being 1860–1862. Fortune is said to have introduced c. 120 plant species to Europe.

[20] Morphett & Tolley 1999, p. 22.
[21] Elliot & Jones 1993, *Encyclopaedia of Australian Plants*, vol. 6.

3 Propagation of citrus

[22] It is interesting to note that Ian Tolley of Tolley's Nursery has developed a system of budding onto Trifoliate rootstock that prevents incompatibility symptoms developing. He insists that the bud be placed 30cm above ground level rather that the traditional 10cm.

5 Pruning and training citrus

[23] *Fruit Tree and Grape Vine Pruning: A Handbook for Fruit and Vine Growers* (1921), Department of Agriculture, South Australia, p. 224.
[24] Ibid.
[25] *The One Straw Revolution* and *The Natural Way of Farming*.

6 Pest and disease control of citrus

[26] McAlpine, D. 1899, *Fungus Disease of Citrus Trees in Australia, and Their Treatment*, Department of Agriculture, Victoria, pp. 74–113.

bibliography

ABC Television, 22 October 1999, Gardening Australia, 'Plastic Surgery' (segment about grafting with Allen Gilbert).

Aitken R. & Looker, M. 2002, *The Oxford Companion to Australian Gardens*, Oxford University Press, Melbourne.

Albrecht, W. A. 1975, *The Albrecht Papers*, vols I–IV, Acres, Missouri, USA.

Alexander, D. McE. 1983, *Some Citrus Species and Varieties in Australia*, CSIRO, Adelaide.

Allen, J. 2002, *Paradise in Your Garden: Smart Permaculture Design*, New Holland Publishers, Sydney.

Australasian Biological Control 1995, *The Good Bug Book*, Australasian Biological Control Inc et al., Queensland.

Australian Citrus Fruit Buyer's Guide 1964, Commonwealth of Australia booklet.

The Macquarie Dictionary of Trees and Shrubs 1986, Macquarie Library, Dee Why, NSW.

Horticulture Australia 1995, Morescope Publishing, Hawthorn East, Vic.

Bailey, F. M. 1909, *Comprehensive Catalogue of Queensland Plants, Both Indigenous and Naturalised*, Queensland Government, Brisbane.

Baker, H. 1998, *The Fruit Garden Displayed* (eighth edition), Royal Horticultural Society, London.

Baxter, P. 1991, *Fruit for Australian Gardens*, Pan Macmillan, Chippendale, NSW.

Beazley, M. 1997, *The Complete Book of Plant Propagation*, Reed International Books Ltd, London.

Bennet, P. 1989, *Australia and New Zealand Organic Gardening*, Child & Associates, French's Forest, NSW.

Bevington, K. B. & Sarooshi, R. A. 1974, *Virus Induced Dwarfing of Citrus*, NSW Department of Agriculture, Dareton.

Boas, I. H. 1947, *The Commercial Timbers of Australia: Their Properties and Use*, Council for Scientific and Industrial Research, Melbourne.

Brock, J. 1988, *Top End Native Plants*, John Brock, Darwin.

Brock, J. 1993, *Native Plants of Northern Australia*, Reed, Chatswood, NSW.

Bryant, G. 2002, *Plant Propagation A to Z: Growing Plants for Free*, Lothian Books, South Melbourne.

Bulford, A. 1999, *Caring for Soil*, Kangaroo Press, Roseville, NSW.

Campbell, S. 1991, *The Mulch Book*, Storey Publishing, Vermont, USA.

Cherikoff, V. & Isaacs J. n.d., *The Bush Food Handbook*, Ti Tree Press, Balmain, NSW.

Clay, R. P. 1972, *Lemons for the Home Garden*, Department of Agriculture Leaflet H227, Victoria.

Clayton, S. 1994, *The Reverse Garbage Mulch Book*, Hyland House, Melbourne.

Coombs, B. ed. 1995, *Horticulture Australia*, Morescope Publishing Pty Ltd, Hawthorn East, Vic.

Cooper, W. & Cooper, W. T. 1994, *Fruits of the Rainforest: A Guide to Fruits in Tropical Rain forests*, Geo Productions Pty Ltd, Sydney.

Cresswell, G. C. & Weir, R. G. 1997, *Plant Nutrient Disorders,* vol. 5: *Ornamental Plants and Shrubs,* Inkata Press, Sydney.

CSIRO 1970, *Insects of Australia,* Melbourne University Press, Melbourne.

Cundall, P. 1989, *The Practical Australian Gardener,* McPhee Gribble (Penguin Books), Ringwood, Vic.

Department of Agriculture 1972, *Pruning of Citrus Trees,* Department of Agriculture Leaflet H115, Victoria.

Despeissis, A. 1921, *The Handbook of Horticulture and Viticulture of Western Australia,* WA Department of Agriculture, Perth

Dunkan, J. H. 1979, *Guidelines for Commercial Citrus Dwarf Plantings,* NSW Department of Agriculture Leaflet H74/67.

Dunkan, J. H. 1980, *Grapefruit Growing,* NSW Department of Agriculture, Agdex 223, Government Printer, NSW.

Edmunds, A. n.d., *Espalier Fruit Trees,* Horticultural Press, Carlton, Vic.

Elliot, W.R. & Jones, D.L. 1982, *Encyclopaedia of Australian Plants Suitable for Cultivation,* vol. 2, Lothian Books, Port Melbourne.

Elliot, W. R. & Jones, D. L. 1993, *Encyclopaedia of Australian Plants Suitable for Cultivation,* vol. 6, Lothian Books, Port Melbourne.

Elliot, W.R. & Jones, D.L. 1997, *Encyclopaedia of Australian Plants Suitable for Cultivation,* vol. 7, Lothian Books, Port Melbourne.

Ellis, B.W. & Bradley, F.M. 1996, *The Gardener's Handbook of Natural Insect and Disease Control,* Rodale Press, Emmaus, Pennsylvania, USA.

Esau, K. 1962, *Anatomy of Seed Plants,* John Wiley & Sons, New York.

Fleming, D. 1992, *Fleming's Deciduous Fruit and Ornamental Trees,* Fleming's Monbulk Nurseries, Vic.

Fowler, C. & Mooney, P. 1990, *Shattering: Food, Politics and the Loss of Genetic Diversity,* The University of Arizona Press, Tucson, USA.

French, C. 1893–1911, *A Handbook of the Destructive Insects of Victoria,* vols 1–5, Government Printer, Melbourne.

French, J. 1995, *Soil Food,* Aird Books, Flemington, Vic.

Fukuoka, M. 1978, *The One Straw Revolution,* Rodale Press, Emmaus, Pennsylvania, USA.

Fukuoka, M. 1985, *The Natural Way of Farming: The Theory and Practice of Green Philosophy,* Japan Publications Inc., New York.

Garner, R. J. 1947, *The Grafter's Handbook,* Cassell Publishers Ltd, London.

Gilbert, A. 1978, *Short Course Notes: Propagation,* unpublished notes, Victorian Department of Agriculture, Melbourne.

Gilbert, A. 1991, *Yates Green Guide to Gardening: A No Fuss Guide to Organic Gardening,* Angus & Robertson/HarperCollins, Pymble, NSW.

Gilbert, A. 1992, *No Garbage: Composting and Recycling,* Lothian Books, Melbourne.

Gilbert, A. 1995, 'Improving Grafting Techniques for Apples', in *International Plant Propagators' Society Combined Proceedings,* vol. 45, pp. 63–7.

Gilbert, A. 2001, *All About Apples,* Hyland House, Melbourne.

Gilbert, A. 2001, *Organic Gardening for the Home Garden* (Yates Mini Guide), HarperCollins, Pymble, NSW.

Gilbert, A. 2001, *Trees and Shrubs for the Home Garden* (Yates Mini Guide), HarperCollins, Pymble, NSW.

Gilbert, A. 2003, *No-Dig Gardening: How to Create an Instant, Low-Maintenance Garden,* ABC Books, Sydney.

Gilbert, A. 2005, *Just Nuts,* Hyland House, Melbourne.

Glowinski, L. 1997, *The Complete Book of Fruit Growing in Australia,* updated paperback edition, Lothian Books, Port Melbourne.

Godden, G. 1978, *Citrus Trees in the Home Garden,* Department of Agriculture Agdex 220/12, Victoria.

Godden. G. 1988, *Growing Citrus Trees,* Lothian Publishing Company Pty Ltd, Melbourne.

Goode, J. & Wilson, C. 1987, *Fruit and Vegetables of the World,* Lothian Publishing Company, Port Melbourne.

Greive, M. 1980, *A Modern Herbal,* Penguin Books, Ringwood, Victoria.

Hamilton, G. 1987, *Successful Organic Gardening,* McMillan, Melbourne.

Hammer, P. R. 1991 *The New Topiary : Imaginative Techniques from Longwood Gardens,* Garden Art Press, UK.

Handreck, K. 1993, *Gardening Down Under,* CSIRO Publications, East Melbourne.

Hely, P. C. et al. 1982, *Insect Pests of Fruit and Vegetables in New South Wales*, Inkata Press, Melbourne.

Hibbert, M. 2004, *The Aussie Plant Finder 2004*, Florilegium, Glebe, NSW.

Holmgren, David 2002, *Permaculture: Principles and Pathways Beyond Sustainability*, Holmgren Design Services, Hepburn, Vic.

Howard, Sir Albert 1943, *An Agricultural Testament*, Oxford University Press, London.

Isaacs, J. 1987, *Bush Food: Aboriginal Food and Herbal Medicine*, Weldons, McMahons Point, NSW.

Janson, H. F. 1996, *Pomona's Harvest: An Illustrated Chronicle of Antiquarian Fruit Literature*, Timber Press, Portland, Oregon, USA.

Jenkins, J. 1999, *The Humanure Handbook*, Jenkins Publishing, Grove City, PA, USA.

Jones, D. L. 1986, *Ornamental Rainforest Plants in Australia*, Reed Books, French's Forest, NSW.

Kains, M. G. & McQuesten, L. M. 1942, *Propagation of Plants*, Orange Judd Publishing Co. Inc., New York.

King, F. H. 1911, *Farmers of Forty Centuries*, Rodale Press, Emmaus, Pennsylvania, USA.

Lake, J. 1996, *Gardening in a Hot Climate*, Lothian, Port Melbourne.

Lavelle C. & M. 2003, *The Organic Garden*, Hermes House, London.

Lord, E. E. 1967, *Shrubs and Trees for Australian Gardens*, Lothian Publishing, Melbourne.

McAlpine, D. 1899, *Fungus Diseases of Citrus Trees in Australia, and Their Treatment*, Government Printer, Melbourne.

McAlpine, D. M. February 1949, 'Winter Drop of Grapefruit: Reduction by Spraying', in *The Journal of the Department of Agriculture*, Victoria.

McCrorey, G. June 1969, 'Making the Most of Citrus', in *The Journal of the Department of Agriculture*, Victoria,.

McDonald, B. 1986, *Practical Woody Plant Propagation For Nursery Growers*, vol. 1, Timber Press, Portland, Oregon, USA.

McEwin, H. 1913, *The Fruitgrower's Handbook*, Commonwealth of Australia, Melbourne.

McLeod, J. 1994, *Heritage Gardening*, Simon & Schuster, East Roseville, NSW.

McMaugh, J. 2001, *What Garden Pest or Disease is That?* New Holland Publishers, Sydney.

McPhee, J. 1966, *Oranges*, Noonday Press, New York.

Malins, J. 1992, *The Essential Pruning Companion*, David & Charles, Devon, UK.

Marshall, T. 2003, *Recycle Your Garden: The Essential Guide to Composting*, ABC Books, Sydney.

Mason, J. 2002, *Propagation from Cuttings*, Kangaroo Press, East Roselle, NSW.

Menzies, G. 2003, *1421: The year China Discovered the World*, Bantam Books, Great Britain.

Mollison, B. 1991, *Introduction to Permaculture*, Tagari Publications, Sydney.

Morphett, B. & Tolley, I. 1999, *Citrus for Everyone*, Adelaide Botanic Handbook, no. 5, Adelaide Botanic Gardens, Adelaide.

Murphy, D. 1993, *Earthworms in Australia*, Hyland House, Melbourne.

New Zealand Biodynamic Association 1989, *Biodynamics: New Directions for Farming and Gardening in New Zealand*, Random House, Albany, New Zealand.

Platt, R. G. 1981, *Micronutrient Deficiencies of Citrus*, Division of Agricultural Services, Leaflet 2115, University of California.

Queensland Department of Primary Industries 1975, *Citrus Bulletin for Queensland Growers*, Queensland.

Queensland Department of Primary Industries 1975, *Citrus Growing in Queensland*, Horticutural Branch, Queensland.

Queensland Department of Primary Industries 1975, *Citrus Growing in the Home Garden*, Department of Primary Industries, Queensland.

Quinn, G. 1921, *Fruit Tree and Grape Vine Pruning: A Handbook for Fruit and Vine Growers*, Government Printer, Adelaide.

Reid, D. P. 1987, *Chinese Herbal Medicine*, CFW Publications, Hong Kong.

Roads, M. J. 1989, *The Natural Magic of Mulch*, Greenhouse Publications, Melbourne.

Rodd, Tony (chief consultant) 1997, *Botanica*, Random House, Milson's Point, NSW.

Rodd, Tony (chief consultant) 1999, *Botanica's Pocket Annuals and Perennials*, Random House, Milson's Point, NSW.

Rodd, Tony (chief consultant) 1999, *Botanica's Pocket Trees and Shrubs*, Random House, Milson's Point, NSW.

Sanders, T. W. 1964, *Sanders' Encyclopaedia of Gardening* (revised by A. G. L. Hellyer), W. H. & L. Collingridge, Florida, USA.

Stephenson, W. A. 1973, *Seaweed in Agriculture and Horticulture*, E. P. Publishing, Yorkshire, UK.

Stern, Y. M. 1993, *Halachos of the Four Species*, Feldheim Publishers, Jerusalem.

Stewart, A. 1999, *Let's Propagate: A Plant Propagation Manual for Australia*, ABC Books, Sydney.

The Staff of The Liberty Hyde Bailey Hortorium 1997, *Hortus Third*, Macmillan Publishing Co. Inc., New York.

The United States Department of Agriculture 1959, *Food: The Yearbook of Agriculture*, The United States Government Printing Office, Washington DC.

Thornton, I. & El-Zeftawi, B. M. 1983, *Culture of Irrigated Citrus Fruits*, Government Printing Service, Victoria.

Usher, G. 1974, *A Dictionary of Plants Used by Man*, Constable and Company, London.

Valder P. 1999, *The Garden Plants of China*, Florilegium, Glebe, NSW.

Victorian Department of Agriculture 1965, *Basis of Success in Citrus Growing*, Victorian Department of Agriculture, Leaflet H.76, Mildura.

Victorian Department of Agriculture 1970, *Mandarin Varieties and Rootstocks*, Victorian Department of Agriculture, Mildura.

Watson, W. (ed.) 1936, *The Gardener's Assistant*, vol. 3, The Gresham Publishing Co., London.

Webber H. J. and Batchelor, L. D. (eds) 1948, *The Citrus Industry*, vols 1 & 2, University of California Press, Berkeley and Los Angeles.

Webber, R. T. J. 1969, *Report: Study Tour Viticultural and Citrus Industries USA*, Department of Agriculture, Mildura, Vic.

Wells, A. et al. 1975, *Citrus Growing in Sunraysia*, Department of Agriculture, Victoria.

Williams, K. A. W. et al. 1992, *The Complete Book of Patio and Container Gardening*, Ward Lock, London.

Williams, K. A. W. 1979, *Native Plants Queensland*, vol. 1, Keith A. W. Williams, North Ipswich, Australia.

Williams, K. A. W. 1984, *Native Plants Queensland*, vol. 2, Keith A. W. Williams, North Ipswich, Australia.

Williams, K. A. W. 1987, *Native Plants Queensland*, vol 3, Keith A. W. Williams, North Ipswich, Australia.

Wilson, E. 1999, *Worm Farm Management*, Kangaroo Press, NSW.

Wilson, E. E. & Ogawa, J. M. 1979, *Fungal, Bacterial and Certain Nonparasitic Diseases of Fruit and Nut Crops in California*, University of California, Los Angeles.

Wishart, R. L. 1974, *Microbudding of Citrus*, Department of Agriculture, Extension Bulletin no. 1874, South Australia.

Woodrow, L. 1996, *The Permaculture Home Garden*, Penguin Books, Ringwood, Vic.

Wright, J. I. 1983, *Plant Propagation for the Amateur Gardener*, Blandford Press, Poole.

Yepsom, R. B. Jr. (ed.) 1996, *Organic Plant Protection*, Rodale Press, Emmaus, Pennsylvania, USA.

Yoshimura, Y. and Halford, G. M. 1957, *The Japanese Art of Miniature Trees and Landscapes*, Charles E. Tuttle Company, Tokyo.

glossary

Acidity/akalinity: soil, water or the taste of a thing is said to be acidic when it has an acid-like consistency or taste, and a low acidity reading. Soil acidity is measured by an acidity/alkalinity pH meter within the 0–14 range. A reading below pH 7 is considered acidic, readings above pH 7 are considered alkaline (see also *pH*).

Aerial layering or marcotting is a plant propagation method based on propagation from a stem or branch piece while the branch is still attached to the tree. It involves cutting the stem or removing a section of bark from around the limb. Sometimes root promoting substances are applied to the injured section to promote the development of roots. The stem can then be wrapped with moist moss, tissue, copra peat, peat moss or some such material and then wrapped in foil or plastic. When roots develop, the wrapping is removed and the branch cut off and planted out in a pot or into the ground.

Alternate bearing (or biennial cropping) is where a tree bears a heavy crop of fruit one year and a much-reduced crop the following year and this continues as a regular pattern.

ACIP – Australian Citrus Improvement Program. ACIP, in consultation with the Australian citrus growers, sources and sponsors desirable imports of citrus budwood. Local cultivars have also been nominated for the program; these are virus indexed and, where possible, viruses are eliminated. Budwood is quarantine indexed at the Elizabeth Macarthur Agricultural Institute (EMAI) in Camden, NSW. Both heat therapy methods and *in vitro* shoot-tip grafting are used to eliminate viruses and viroids often in conjunction with each other. Virus-free true-to-type stock (two trees of each variety) is maintained in insect-proof screenhouses. The program aims to supply clean budwood of high performance clones (cultivars) to the citrus industry in the shortest possible time. Field- and greenhouse-grown budwood source trees are located at the Agricultural Research and Advisory Station at Dareton, NSW, and a supplementary supply of new varieties is grown at Monash in South Australia managed by the

Injured lemon tree trunk that was coated with clay, initiating root growth, see also, page 44.

South Australian Citrus Improvement Society (SACIS). Rootstock seed supply trees are grown at Dareton and Gosford in NSW, and Monash in South Australia, and the Queensland Citrus Improvement Society also runs a citrus seed supply scheme. See also *Auscitrus*.

Auscitrus is the trading name of the Australian Citrus Propagation Association Incorporated (ACPAI), and is an initiative of the Australian Citrus Propagation Association (ACPA), the Australian Citrus Improvement Association (ACIA), the Horticultural Research and Development Corporation (HRDC), and NSW Agriculture. The Auscitrus program is funded through the sale of élite buds and a levy on seed and budwood sales from State multiplication schemes matched in dollar terms by the HRDC. Recently imported cultivars are performance evaluated before release; this can take from 1–4 years. Both Government agencies and a group of nominated nursery operators are involved. Bulking up occurs in field multiplication sites, for scions and rootstocks, by the various State associations and then sold on to growers and plant nurseries.

Basal flooding or *Basal irrigation* means watering upward from the base of a plant. One way this can be done is by placing the pot or container in a trough or large dish to allow take-up of moisture through the roots (this is in comparison to watering a plant from above).

Biennial cropping: see *Alternate bearing*.

Biodynamic growing is an anthroposophical gardening and farming method based on the teachings of the Austrian social philosopher Rudolph Steiner (1861–1925). The approach includes concepts of life forces and the effects of planetary bodies on the growth of plants and animals.

Bordeaux A mixture of copper sulphate and lime used mainly to control bacterial diseases.

Bordeaux paste is made by mixing Bordeaux powder (see above) with water and a few drops of olive oil or petroleum oil to make the paste stick. It is used for painting wounds and pruning cuts to prevent disease infection.

Budding is the act of placing a bud under the bark of a rootstock tree to produce a tree or branch bearing a selected cultivar.

Bud union is that point where a bud has been inserted or a scion grafted to the rootstock plant and usually shows as a change in bark characteristics such as a slight swelling or a bending of the trunk, or a scarred tissue area. It is usually about 10–30cm above ground level on standard nursery trees.

Callusing is a healing growth that covers wounds.

Calyx is the outer whorl of floral envelopes showing as a green button at the stalk end of citrus fruit.

Cambium layer is a layer of indeterminate cells just under the bark layer of a trunk or limb of a tree. It is where healing tissue necessary for successful grafting is situated.

Chamfer means to cut a sharp edge at an angle, thus creating a sloped, cut edge. On a cut limb this is done to bark level to lessen the pressure of the bark and allow easier callusing to occur on its end.

Chimera is when citrus fruit shows two different types of skin growth on the one fruit.

Chip-budding refers to the placing of a chip-shaped bud from one plant to another (rootstock) to propagate a particular cultivar.

Chunk pruning means pruning out chunks of the tree canopy to allow better light penetration.

Cincturing is deliberate removal of a strip of bark around a stem or branch used for propagation purposes to initiate roots at the injury site or to initiate fruiting (see also *Aerial layering*).

Companion planting: by planting bee-attracting plants to ensure pollination, attractive flowering plants providing food pollen for birds, butterflies, wasps, hover flies and other predators that parasitise or eat pest insects, fruiting trees are kept healthy. Companion plants can also provide shade, supply extra nutrients, give off scents that repel insect pests and aid plant growth.

Composting is an easy process and there are many ways of composting. Large compost heaps using hay bales or other forms of enclosure can be used in larger gardens. Mature compost heaps can easily be converted to non-dig gardens or the compost can be used in pots or containers or spread over existing gardens. Wherever compost is used it attracts worms that, in turn, aerate soil and move compost material to where other soil organisms such as bacteria and fungi can feed on it creating humus (see also *Humus*).

Copper sprays are copper-based materials used for disease control (see also *Bordeaux* and *Bordeaux paste*).

GLOSSARY | 161

Cross-pollination is when pollen from one tree pollinates the flowers of another tree. Pollen can be transferred by wind, insects or bees. All citrus are self-pollinating but will accept pollen from other cultivars and species (see also *Self-pollination*).

Cultivar is the name given to a plant variety that has been deliberately selected or bred using various plant breeding techniques. Particular cultivars can be named and registered under PVR (Plant Variety Rights) and breeders can obtain royalties for buds, scions or trees through PBR (Plant Breeders' Rights).

Defoliation is extensive and sometimes complete loss of leaves and usually occurs as the result of a major shock to the tree's system. Such shocks can include waterlogging, over fertilisation, heavy frost or snow. Severe pest damage (see also *Leaf drop*).

Etiolation means excluding or reducing the light to plant growth to increase its ability to propagate successfully.

Fertigation refers to using irrigation water to introduce soluble fertilisers, i.e. irrigation and fertiliser application occur at the same time.

Fertiliser is any substance, either natural (i.e. organic) or human-made (i.e. chemical) added to soil or other growing medium to aid in plant growth by supplying additional nutrients. Manure, for example, is a natural fertiliser. Fertilisers can now also be applied to foliage. See also *Foliar fertiliser*.

Flesh (pulp) refers to the juicy or edible part of citrus fruit that is enclosed by skin (rind).

Foliar fertiliser is a fertiliser used in a liquid form and applied by spraying onto plant foliage

Forcing is a term used to describe a situation where plants are grown in an artificial environment such as a controlled atmosphere greenhouse under ideal conditions for growth, thus forcing the plant to grow

Frass is a material like wood sawdust left as stem boring larvae burrow into wood or the material excreted by insects as they digest plant matter. Gamma, see *Plant nursing*

Girdling occurs when insects remove bark right around the circumference of a twig or limb, thus effectively ring-barking the twig or branch.

Grafting is the transfer of a bud or scion from a selected cultivar to a chosen rootstock tree

Hesperidium is the botanical term for the citrus fruit.

Honeydew is the name given to secretions of a sugary by-product from feeding insects such as scale and aphids.

Humus is a rich organic food, created from the breakdown of compost or organic matter. It is a gel-like substance that feeds plants and acts like an organic glue holding particles of soil together improving soil structure.

Hybrid: A hybrid is created when two different species are used for pollination purposes resulting in an entirely different plant.

Integrated control measures are the use of a combination of biological and chemical measures to control pests and diseases. Generally less chemicals are used in integrated approaches.

Integrated Pest Management (IPM), see *Integrated control measures*.

Internode is the name given to the distance between one leaf and the next on a growth lateral.

Laterals are short new shoots, longer than 10cm, that grow in new growth flushes from around pruning sites and from all over the tree. When laterals are removed from the tree, others will grow from the base of the cut area. If laterals are cut in half they can produce flower buds and shoots. In most cases they just produce a group of more long laterals.

Leaching (of soil or potting media) happens when water percolates through soil, dissolving salts and nutrients and carrying them to the subsoil or into the water table where they are unavailable to plants.

Leaf drop is when leaves drop from the tree, and can occur naturally as a result of leaf ageing, or may be a symptom of disease, poor nutrition, wind or frost damage or other environmental factors (see also *Defoliation*).

Marcotting, see *Aerial layering*

Micronutrients are elements other than the principal ones (nitrogen, phosphorus, potassium and calcium) that are required for plant health. These include copper, magnesium, manganese, sulphur and zinc.

Monoembryonic describes a seed containing a single embryo.

Multi grafting is the placing of many grafts on one rootstock or tree. Often the individual grafts are of different cultivars.

Natural drop of flowers, and of developing or maturing fruit occurs regularly in citrus. It is a natural phenomenon caused by a natural thinning process within the tree, or it can be caused by environmental conditions or nutritional deficiencies.

Nuclei refers to embryos within a cell or seed.

NPK ratio refers to the Nitrogen (N), Phosphorus (P) and Potassium (K) ratio in a fertiliser mix.

Nucellar are nuclei within the seed of some citrus, for example the lemon, that are not the result of cross-pollination. These nuclei will produce plants and fruit exactly the same as the parent plant from which the seed was obtained. Nucellar plants are usually slower to fruit, can produce thorny growth and are relatively free of virus infection. Selections of these nucellar trees are used for rootstock material or, if they produce good fruit, for planting out into citrus groves. See also *Polyembryonic*.

Organic gardening is simply gardening done using natural processes and without the use of synthetically manufactured chemicals such as weedicides, herbicides or artificial fertilisers. It is generally also associated with an approach to living that sees the world as an interconnected system that necessitates gardening being environmentally friendly and safe for humans as well as for other species. Organic gardening is related to permaculture and biodynamic gardening – all use methods that are in tune with nature and natural processes, emphasising, in their different ways, growing healthy plants while improving soil structure and enhancing quality of life.

Organic fertilisers include animal manures (chicken, sheep, cow), liquid supplements such as my partner's favourite, liquefied comfrey, and fish and seaweed supplements. There are many registered organic materials available: fish waste, natural rock nutrients, seaweed products and packaged compost.

Organic groups: There are many large registered organic groups across Australia and many local organic gardening clubs and associations where further information can be obtained. Organic gardeners are usually interested in saving rare and heritage cultivars of plants, fruit and vegetables and can also join seed savers networks to enable them to source rare seeds or plant material.

Organic mulches include shredded paper, hay, wheat straw, pea straw, baled lucerne, lupin straw, bagasse, sawdust, bark and wood chips, compost and lawn clippings. Green grass or weeds can be mulch-mown or cut with a whipper snipper to give green mulch cover. Organic gardening mimics processes that occur in nature, so deep mulch layers on soil mimic the way leaf litter naturally mulches soil. Mulches help keep plant roots cool in hotter weather, retain moisture thereby reducing water usage, encourage soil microorganisms, and, as they break down, further feed soil. Mulches used over layers of wetted newspaper can help suppress weeds, reducing time spent on weeding, and increasing time available for smelling the roses. Supplementary organic fertilisers such as blood and bone can be used to help the mulch material break down quickly.

Organic practices include out crop rotation of vegetable crops to prevent build-up of diseases; plant vegetables only when their major pests are not present, and use companion planting to attract beneficial insects and insectivorous birds. Organic gardeners often use no-dig gardens because they are low impact, help improve soils and are good for recycling organic waste. Recycling plant and other organic materials to form rich compost is one of the basics of organic gardening; compost feeds the soil and, thus, the plants growing in it. Organic practices include such natural approaches to pest and disease control as use of pyrethrum, sulphur and garlic sprays for insect control as well as predator insects.

Organic standards: There are some specific principles and requirements associated with marketing food and other products labelled as 'organic'. The Australian Quarantine Service (AQIS) administers National Standards for Organic and Biodynamic Produce, and licenses a number of organisations to carry out certification processes. Food and other products sold as 'organic' must be certified.

Ovipositor is the egg laying apparatus of female insects.

Parthenocarpic describes fruit, such as some navel oranges, in that they produce fruit without any seed.

Parthenogenetic means producing young or fruit without fertilisation.

PBR (Plant Breeders' Rights) is the name given to the right of plant breeders to obtain royalties from plant material they have bred and registered. For example, a person who has developed a new nut cultivar by cross-breeding receives royalties under PBR every time the plant is propagated for sale. See also PVR.

Pectin is a fruit extract used to help 'set' jam or jellies. Some fruit, such as apples, have a very high natural pectin content while others, such as strawberries are relatively low in pectin.

Pectin test for marmalade: pectin content can be tested by adding 3 tsp of methylated spirits to 1 tsp of stirred cold mixture. Stir gently and leave for 1–2 minutes. If the result is one good clot, the pectin content is excellent and the ratio of sugar can be 1:1 cups. If the result is 2–4 smaller clots, the pectin content is moderate and $3/4$ cup sugar to 1 cup mix will produce a better result. If there is no clotting or many small clots, it may be necessary to boil mixture longer or add additional pectin such as Jamsetta ™ (this may be preferable to grossly overboiled fruit and will not affect taste).

Permaculture refers to the concept of permanent agriculture developed by Australians David Holmgren and Bill Mollison. Permaculture is an integrated approach to farming, gardening, animal husbandry, fish culture, horticulture and human activities.

Photosynthesis is the process in plants of synthesising carbon dioxide and water into food (carbohydrates) using chlorophyll in their leaves and sunlight or light as energy and creating oxygen as a by-product.

pH is a measure of acidity or alkalinity. In the range pH 0–14, pH 7.0 is neutral, pH 3.0 is low acid and pH 10 is highly alkaline.

Pitam is a persistent stylar growth at the base of the fruit.

Plant naming, or nomenclature, is based on a uniform system that owes much of its origins to the work of the great Swedish naturalist, Linnaeus (1707–78). The angiosperms, the major flowering plant group on earth, are divided into two main groups, the dicotyledons (having two *cotyledons* or seed leaves) and the monocotyledons (having one cotyledon). Both groups are then further subdivided into orders, families, genera and species, with a species being a group of plants with the closest similarities and the most distinct and very specific characteristics. Genera are groups of related species while families are groups of related genera and so on for orders of plants. The genus *Citrus*, for example, contains all the species of citrus including limes, lemons, oranges, citrons, grapefruit, mandarins and finger limes. A species within this group is, for example, the citron *Citrus medica*.

Naming conventions are quite strict. Each species has a common name (or a number of common names) and a botanical name and, while common names may vary, the botanical name generally does not (although a species can have synonymous names, and names can be changed when, for example, new information leads to a reclassification of a species). The botanical name for each species consists of a combination of its genus and species names, e.g. the citron is identified botanically as *Citrus medica* with *Citrus* being the genus name and *medica* being the species name.

How botanical names are written also has strict rules. They are usually written in italics, with the genus name first and capitalised, the species name second and not capitalised. A species name can also have a subspecies included (e.g. the blue gum *Eucalyptus globulus* ssp. *globulus*) or a species can have a variety (named cultivar) name attached to it as is common in the case of such species as roses (e.g. *Rosa roxburghii* 'Plena'). To complicate naming conventions, variety names are not italicised and are enclosed in inverted commas. There are many other rules and conventions but these are the most common.

These conventions are not always well adhered to and this can lead to confusion. It is important to use correct botanical names wherever possible to make sure that plants are identified correctly.

Polyembryonic: A polyembryonic seed contains many nuclei. Some nuclei within the seed result from cross-pollination with other citrus cultivars or species and have the potential to produce trees with a new type of citrus fruit. See also *Nucellar*.

Proboscis is a hollow needle-like part on some insects' mouth-parts used for sucking.

Pulp, see *Flesh.*

PVR. (Plant Variety Rights) allows plant breeders to register newly bred plants and to receive royalties from the sale of propagation material of that plant. See also PBR.

Rootstock is a plant onto which a known variety or cultivar is budded or grafted and it provides the roots for the budded or grafted tree. Often particular rootstocks have distinct advantages such as being disease-resistant, having a dwarfing effect, or being capable of forming a better root system for the grafted plant.

Rosetting is when bunched growth with many growth shoots with short internodes (see *Internode*) occurs.

Scions are small, round-stemmed lateral pieces containing 3–6 buds taken from a fruitful part of the donor tree. Leaves are usually removed but one or more can be left attached to the top of the scion providing that a plastic sleeve is placed over the scion after it has been inserted as a graft.

Scurvy is a medical condition characterised by general debility of the body, foul breath, tender gums, sub-cutaneous eruptions and painful limbs caused by lack of Vitamin C.

Seaweed, liquid solution is made up using concentrated liquid seaweed extract and is used as a foliar spray and for root drenching.

Seedling variation refers to the fact that plants, if grown from seed, can produce hundreds of plants similar to the parent plants, but many, although genetically similar, produce fruit that are entirely different in size, colour, taste, or ripening habit. In other words they do not grow 'true to type'. Seedling variations may be chosen as new cultivars.

Self-pollination refers to pollen transfer between flowers of a particular cultivar on the same tree. All citrus are self-pollinating but will accept pollen from other cultivars and species (see also *Cross-pollination*).

Setting point for marmalade (from Norma Campbell) is when marmalade mixture jells and is ready to bottle. Test by placing approximately 1 tsp on a plate and putting in freezer for about 10 minutes. If it then crinkles when pushed with finger, it is ready.

Skirting is the removal of the lower foliage of the citrus tree canopy to reduce dust and to improve air movement through the tree, thus reducing the likelihood of diseases and access to such pests as slugs and snails.

Sport – A naturally occurring *mutation* on a tree or a shoot or bud that gives rise to a different characteristic, and often to spur-forming varieties of entirely new varieties.

Sterilising jars for marmalade or jam is essential. Wash jars thoroughly and rinse in hot water. Place in a pot of boiling water for 10 minutes and then drain and dry in an oven set at 150°. Remove jars from oven just before using and fill while hot.

Thinning of fruit, reducing the number of fruit on a tree, is necessary to increase the size of the remaining fruit.

Trellis refers to a support structure upon which trees can be trained or tied to obtain a set design shape.

Trifoliata is a species (*Poncirus trifoliata*) related to citrus used as rootstock.

Trifoliate leaves have three leaflets per leaf.

TSS is the abbreviation for Total Soluble Solids, which is one of the standards used for citrus fruit maturity used by commercial growers (see also *TTA*).

TTA is the abbreviation for Total Titratable Acidity, which is one of the standards used for citrus fruit maturity used by commercial growers (see also *TSS*).

Topiary is a form of pruning and shaping trees into geometric forms or other shapes such as animals or birds.

Virus refers to microscopic disease organisms.

Virus indexing is a technique for assessing the presence of particular viruses in plant material. Sections of a plant are grafted onto another that is sensitive to known viruses; the virus, if present, will show on the indexing plant.

Watershoots are very strong growing shoots that can grow a metre or more during one season. They often grow from cut branches or from areas around the central base of the tree.

Whitewash is a powdered lime and water mix used as a paste or spray for plant protection.

Wound dressing paint is used for covering wounds or pruning or saw cuts on tree limbs or branches. A good wound dressing paint can be made from Bordeaux powder mixed to a paste with a little water and a few drops of oil to help the paint stick. This is used to paint the cut surfaces of pruned branches. Alternatively, a white, water-based paint can be applied to the cut surface area, but do not use oil-based paints as these will poison the tree.

index

acidity/alkalinity, *see* soil
ACIP (Australian Citrus Improvement Program), *see* standards, citrus industry
aerial layering, 44–5, 159
air flow, disease prevention, 83, 117, 120, 139, 141
alternate bearing, *see* biennial cropping/biennial bearing
Auscitrus, *see* standards, citrus industry
Australian citrus species, 1, 2, 9, 14, 15, 16, 40–1, 87, 101

bark graft, 42, 59–60
biennial cropping/biennial bearing, 23, 37, 80. 82, 97, 125, 159, 160
birds
   as pests, 91–141
   pest predator, 80, 91–141, 160
bonsai, 74, 83, 87
boron, 68, 95, 96, 127, 132–3
bud, 7, 13, 44, 47–8, 52–8, 160, 161, 164
budding, 9, 42, 44, 47–8, 50, 52, 53–63, 84, 120, 128, 140, 154, 160
   bud stick(s), 52, 54, 57
   chip, 42, 48, 53, 55–8
   micro-budding, 57–8
   'T' bud method, 56, 57, 53–5, 58

calamondin, 11, 16, 25, 32–3, 40
calcium, 66, 79, 96, 129, 131, 133, 161
callus/callusing/callused, 48, 53, 56, 60, 63, 72, 100, 116, 160
calyx, 144, 160
cambium layer, 49, 52, 55, 56, 61, 62, 63, 72, 100, 116, 160
chamfer/chamfered, 59, 72, 160, 123
chimera, 60, 160
Chinotto, 16, 25, 33
chip bud, 42, 48, 53, 55–8
chunk pruning, 97, 160
cincturing, 44, 160
citrange, *see* lemon, rough
citron, 5, 16, 25, 28–9, 30, 65, 163
citrus
   Australian, *see* Australian citrus species

genus, 1, 14, 40, 163
groves, 3, 4, 5, 7, 8, 9, 20, 50, 98, 125, 162
   in Australia, 5–10, 13–14, 16, 17, 20, 21, 25, 26, 29, 32, 35, 38, 49, 65, 87, 91, 116
   organic citrus grove, 78–81
products, 11, 140, 144, 146
skins, 80, 146
citrus trees, 82–90
   in pots, 5, 11, 27, 33, 36, 43, 44, 57, 62, 64, 74–7, 84, 89, 128, 132, 133, 134, 137, 160
   open grown, 64–73
   repotting, 75–7
   shifting, 71–3
   skeletonising, 84–5
cleft/wedge graft, 60–1
Clementine, *see* mandarin
cobalt, 80
commercial orchards, 48, 106
companion planting, 99, 113, 160, 162
compost/composting, 43, 67, 79, 80, 102, 111, 131, 134, 139, 140, 146, 160, 161, 162
copper, 68, 79, 92, 95, 96, 129, 131, 133, 161
copper spray, as disease control, 85, 117, 118, 119, 121, 125, 139, 160
cross-pollination, *see* pollination
cultivar(s) (variety), 12–41, 44, 57, 58, 60, 64, 142, 161, 163, 164
names/naming, *see* plant naming
cumquat, 2, 10, 11, 25, 32–5, 41, 53, 64, 72, 74, 76, 85, 87, 105, 126, 143, 144–45, 150
cuttings, 6, 28, 42, 45–7, 52
   leaf bud, 45–6
   semi-hardwood, 46
   softwood, 46
   root, 46–7

defoliation, 120, 137, 161 *see also* leaf drop
dwarfing
   effects, 50
   rootstock, 50, 164

equipment
   budding and grafting, 48–9, 55
   pruning, 83–4

165

espalier, 83, 87–90
  clipped hedge, 89–90
  multiple-'T' shape, 89
  open fan-shape, 89
  pyramid, 89
etiolation, 52, 161

fertigation, 69, 81, 161 *see also* water
fertiliser, 64, 66–8, 69, 81, 129, 130, 134, 137, 138, 161, 162
  foliar 68, 138, 161
  organic 66, 67, 84, 135, 162
flesh, 2, 11, 14, 21, 22, 24, 31, 32, 35, 36, 37, 38, 39, 92, 95, 96, 127, 129, 132, 143, 161
flower drop, 97 126
foliar, nutrient, 68, 69
forcing, 4, 62
frass, 96, 102, 114, 161
frost, 6, 9, 64, 65, 66, 73, 85, 91–141, 161

genera, *see* plant naming
grafting, 4, 42, 44, 45, 47–63, 83, 84, 86, 120, 127, 140, 159, 160, 161
  bark, 42, 59–60
  cleft/wedge graft, 60–1
  equipment, *see* equipment, budding and grafting
  in-vitro grafting, 42
  multi, 47, 161
  scion grafts, 58
  soft-shoot cleft graft, 62
  whip and tongue, 62–3
grapefruit 7, 11, 15, 16, 21, 23, 32, 36–9, 55, 100, 119, 123, 125, 126, 129, 133, 138, 142, 144, 146, 148, 149, 164
greenhouse, 4, 9, 48, 161

hesperidium, 14, 161
honeydew, 80, 98, 99, 101, 110, 11, 126, 161
humus, 139, 160, 161
hybrid, 11, 22, 23, 24, 25, 27, 30, 36, 37, 38, 40, 161
hygiene, 57, 140

iron, 66, 131, 134, 135, 137
irrigation, *see* water

Kaffir lime, 14, 15, 25, 39–40, 41

laterals, 47, 57, 82, 133, 161
layering, 42
  aerial, 44–5, 159
  marcotting, 44, 161
leaf drop, 95, 96, 97, 120, 128, 129, 136, 137, 161 *see also* defoliation
leaching, 66, 136, 161
lemon, 3, 6, 15, 25–8, 41, 51, 55, 64, 70, 71, 73, 74, 82, 86, 87, 91, 92, 96, 98, 100, 109, 116, 118, 119, 121, 123, 128, 129, 134, 135, 136, 137, 144, 146, 147, 149, 159, 162
  Rough, 15, 25, 30, 50, 124
  Meyer, 9, 15, 16, 27, 87, 103, 110

lime, 3, 4, 6, 11, 15, 25, 35–6, 43, 87, 121, 127, 142, 144, 146, 147, 149, 163
  agricultural, 73, 109, 121, 129, 133, 134, 135, 139, 140, 160, 164
  Kaffir, 14, 15, 25, 39–40, 41
  Leech lime, *see* Kaffir lime
  Makrut lime, *see* Kaffir lime (s)
  Musk lime, *see* Calamondin
  Tahitian, 15, 35, 125, 136
  Rangpur, 15, 36, 50, 51
  sweet, 15
  West Indian or Mexican lime, 35

magnesium, 66, 68, 79, 95, 127, 131, 134–5, 161
mandarin, 1, 2, 7, 11, 14–17, 20–4, 32, 36, 40, 44, 50, 51, 56, 60, 74, 78, 80, 81, 83, 85, 87, 95, 98, 110, 113, 117, 119, 121, 123, 126, 142–43, 146, 148, 163
  Clementine, 15, 20, 22, 24
  Satsuma, 15, 20, 24
  Tangelo, 15, 21, 23, 24
  Tangerine, 15, 20
  Tangor, 15, 20, 22, 24
manganese, 66, 68, 79, 95, 131, 135, 161
marcotting, 44, 161
marketing, 8, 162
maturity, 29, 43, 66, 80, 126, 127, 143, 164
micro-budding, 57–8
micronutrients, 66, 73, 161
molybdenum, 80, 131
monoembryonic, 43, 161
mulch, 67, 70–1, 79, 102, 121, 122, 131, 139, 140, 162
  mowing, 79, 140, 162

nitrogen, 64, 66, 67, 95, 96, 128, 129, 131, 136, 161, 162
NPK, *see* fertilisers
nucellar, 43, 162 *see also* monoembryonic, polyembryonic, nuclei
nuclei, 33, 162 *see also* monoembryonic, polyembryonic, nucellar
nutrients, 15, 67, 69, 75, 79, 92, 93, 94, 96, 129, 130–138, 139, 141, 160, 161, 162
  boron, 68, 95, 96, 127, 132–3
  calcium, 66, 79, 96, 129, 131, 133, 161
  cobalt, 80
  copper, 68, 79, 92, 95, 96, 129, 131, 133, 161
  foliar, 68, 69
  iron, 66, 131, 134, 135, 137
  magnesium, 66, 68, 79, 95, 127, 131, 134–5, 161
  manganese, 66, 68, 79, 95, 131, 135, 161
  micronutrients, 66, 73, 161
  molybdenum, 80, 131
  nitrogen, 64, 66, 67, 95, 96, 128, 129, 131, 136, 161, 162
  phosphorus, 66, 67, 96, 131, 136, 161, 162
  potassium, 66, 67, 79, 95, 128, 131, 134, 136–7, 161, 162
  sulphur, 66, 92, 95, 131, 137, 140, 161
  zinc, 66, 68, 79, 95, 131, 135, 138, 161
nutritional deficiencies, *see* nutrients

orange, 3, 4, 7, 8, 15, 27, 72, 81, 83
    Bergamot, 32
    blood, pink or red fleshed, 21, 59
    groves, 5, 78–81
    Navel, 7, 8, 12–13, 17–19, 92, 124
    Panama, see calamondin
    Poorman, 32, 37
    Seville or sour, 3, 11, 30–2, 44, 51
    sweet, 16–17, 51, 59, 80, 108, 11, 121, 123, 126
    Trifoliate, 14, 35, 43, 44, 49, 50 164 see also rootstocks, Trifoliata
    Valencia, 20, 78, 92, 106, 127

orangeries, 4–5
organic fertiliser, see fertiliser, organic
organic standards, see standards, organic
organics/organic gardening/organic practices, ix, 44, 64, 65, 66, 67, 68, 78–81, 84, 131, 132, 135, 136, 137, 151, 161, 162
ovipositor, 103, 106, 162

parthenocarpic, 2, 40, 163
parthenogenetic, 112, 163
PBR (Plant Breeders' Rights), 8, 14, 18, 41, 161, 162–3 see also PVR
pectin/pectin test, 146, 148, 163
permaculture, 106, 163
pH, see soil
phosphorus, 66, 67, 96, 131, 136, 161, 162
photosynthesis, 101, 121, 163
pitam, 130, 163
planting, ix, 44, 64–6, 82, 87, 162
plant naming, 1, 13, 15, 33, 161, 163
plastic sleeve, use in propagation, 48–9, 52, 56, 57, 60, 164
pollination, 43, 127, 160, 161, 162
polyembryonic, 33, 43, see also monoembryonic, nucellar, nuclei
potassium, 66, 67, 79, 95, 128, 131, 134, 136–7, 161, 162
proboscis, 163
pruning, ix, 22, 37, 57, 64, 75, 82–90, 160, see also chunk pruning
    Allen Gilbert's new method of pruning lemons, 86
pulp, see flesh
pummelo, 15, 36, 38–9, 40, 44
PVR (Plant Variety Rights), 8, 13, 18, see also PBR

repotting trees, 75–7
Research, 7–10
rootstocks, 6, 7, 9, 16, 30, 32, 35, 41–2, 44, 46, 49, 51, 52–3, 56, 62, 80, 121, 133, 160, 163–4
    Trifoliata, 14, 23, 32, 37, 43, 50–2, 121, 164, see also orange, Trifoliate
rosetting, 108, 164

satsuma, see mandarin
scion, 45, 49, 51, 52–3, 120, 130, 160, 161, 164 see also budding and grafting
scurvy, 3, 6, 164

seaweed, 5, 66, 68, 72, 73, 75, 77, 79, 84, 87, 127, 131, 141, 162, 164
seed, 2, 3, 5, 12, 13–14, 16, 21, 30, 33, 39, 40, 42–44, 50, 81, 140, 160, 161, 162
seedling, 3, 5, 7, 16, 21, 25, 37, 41, 43, 44, 45, 52, 55, 57, 58, 62, 113, 130
    variation, 2, 16, 41, 43, 164
self-pollination, see pollination
Shaddock, 36, 38, 39
shifting trees, 71–3
skeletonising trees, 84–5
soil
    acid/acidic, 67, 74, 131, 133, 134, 135, 136, 138, 159, 163
    alkaline/alkalinity, 51, 67, 68, 93, 97, 129, 130, 131, 133, 134, 135, 138, 159, 163
sport(s), 7, 12, 13, 16, 17, 18, 19, 23, 25, 37, 164
standards
    citrus industry, 13–14, 163
    organic, 139, 162
sterilising, jars, 164
storage, 16, 143, 144, 145
    cool, 144
    seed and seedlings, 43, 45
sulphur, 66, 92, 95, 131, 137, 140, 161
    as disease control, 109, 162

'T' bud method, 56, 57, 53–5, 58
Tangelo, see mandarins
Tangerine, see mandarins
Tangor, see mandarins
thinning fruit, 24, 80, 83, 110, 164
tissue culture, 42
topiary, 83, 88, 89, 164
Trifoliata, see rootstocks, Trifoliata and orange, Trifoliate
trellis, 87, 164
TSS (Total Soluble Solids), 143, 164 see also standards, citrus industry
TTA (Total Titratable Acidity), 143, 164 see also standards, citrus industry

variety, see cultivar
virus, 51, 53, 62, 81, 99, 127, 128, 159, 162, 164
    indexing 62, 164

water/watering, 7, 15, 54, 64, 68, 69–70, 72, 75, 77, 81, 87, 96, 97, 119, 123, 124, 126, 127, 128, 129
    basal, 139
    irrigation, 7, 120, 129, 132 see also fertigation
waterlogging, 130
wind clipping, 114, 164
weeds, 69, 70, 78, 79, 107, 116, 117, 122, 139, 140, 141, 162
whip and tongue graft, 62–3
whitewash, 72, 84, 164
worms, 139, 141, 160
wound dressing paint, 72, 84, 164

zinc, 66, 68, 79, 95, 131, 135, 138, 161

OTHER BOOKS BY ALLEN GILBERT

ESPALIER

FULL-COLOUR, SEWN FLEXI-COVER
144 PP, 232 X 152 MM
ISBN 9781864471090
$29.95

His unique but simple methods of training plants as espaliers are both original and brilliant. (from Peter Cundall's Foreword)

Espalier covers:
- forms, trellis systems and designs
- selection – including flowering plants, fruit trees, and native Australian plants
- shaping, pruning, maintenance and care

JUST NUTS

FULL-COLOUR, SEWN PAPERBACK
144 PAGES, 232 X 152 MM
ISBN 9781864470918
$29.95

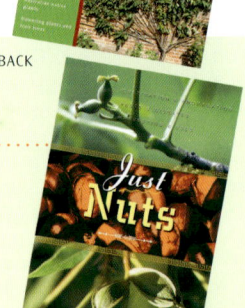

We can all enjoy the abundant and versatile produce of nut trees by growing them ourselves. Whether in the home garden, a hobby farm or commercial orchard, Just Nuts will make it easy. It's a comprehensive guide to:
- tree nuts, their cultivation, harvesting and use
- propagation, including Allen Gilbert's budding and grafting methods
- Allen Gilbert's pruning methods
- a selection of other popular nuts

ALL ABOUT APPLES

FULL-COLOUR, SEWN PAPERBACK
144 PAGES, 245 X 170 MM
ISBN 9781864470468
$31.95

A reliable and informative guide to apples old and new, their care, control of pests and disease, propagation and harvesting.
- Includes a chapter on small-scale orcharding
- Espalier apple trees for unusual landscape effects
- Grow more than ten different apple varieties on a single tree
- Discover the beautiful taste of organic apples in many varieties from yesteryear
- More apples and less work with Allen Gilbert's pruning system

CLIMBERS AND CREEPERS

FULL-COLOUR, SEWN PAPERBACK
128 PAGES, 235 X 155 MM
ISBN 9781864470734
$19.95

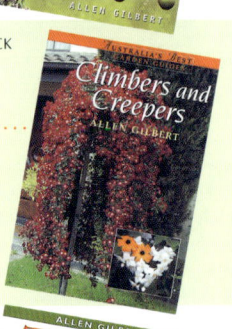

Let the colours, scents, and fruit of these magnificent, rambling, scrambling, crawling and sprawling plants dress up your garden!
- Clear descriptions of the best climbers and creepers
- Good colour photographs of every cultivar, so you can see how the plants you buy will look in your garden
- Australia-wide cultivation notes
- No more aimless trudging through garden nurseries

TOMATOES FOR EVERYONE
A Practical Guide to Growing Tomatoes All Year Round

FULL-COLOUR, SEWN PAPERBACK
152 PAGES, 250 X 185 MM
ISBN 9781864470192
$31.95

A book for people who don't have a lot of space, time or energy, but love the taste of home-grown fruit. Share your garden with Saucy Sue, Mama's Delight, Daydream or Best of All – just a few of the many heritage tomatoes Allen shows you how to grow. Choose from hundreds of varieties, and grow them from seeds or grafts, in non-dig gardens or pots, organically or hydroponically: there's something here for everyone!